11th European Conference on High-Technology Plasma Processes 2010

(HTPP 11)

AA009061

Journal of Physics: Conference Series Volume 275

Brussels, Belgium
27 June – 2 July 2010

ISBN: 978-1-61782-312-1
ISSN: 1742-6588

Printed from e-media with permission by:

Curran Associates, Inc.
57 Morehouse Lane
Red Hook, NY 12571

Printed by Curran Associates, Inc. (2011)

For permission requests, please contact the Institute of Physics
at the address below.

Institute of Physics
Dirac House, Temple Back
Bristol BS1 6BE UK

Phone: 44 1 17 929 7481
Fax: 44 1 17 920 0979

techtracking@iop.org

Additional copies of this publication are available from:

Curran Associates, Inc.
57 Morehouse Lane
Red Hook, NY 12571 USA
Phone: 845-758-0400
Fax: 845-758-2634
Email: curran@proceedings.com
Web: www.proceedings.com

TABLE OF CONTENTS

PREFACE

It is with pleasure and pride that we present the selected contributions from participants of the 11th High-Tech Plasma Processes conference.

This conference, which took place in Brussels from June 28 to July 2 2010, is based on a European forum with a history of more than twenty years. The conference series started as a thermal plasma conference and gradually expanded to include other topics and fields as well. HTPP 11 was organized in collaboration with the Belgian Interuniversity Attraction Pole (IAP): Physical chemistry of Plasma–surface Interactions (PSI–ψ). The program was devised by the plasma group of the Technische Universiteit Eindhoven in collaboration with the IAP, the Association Arc Electrique and the International Scientific Committee. The organization was guided by the Steering Committee and supervised by the two founding members, Jacques Amouroux and Pierre Fauchais.

HTPP aims to bring together different scientific communities to facilitate contacts between science, technology and industry, providing a platform for the exploration of elementary processes in and by plasmas. This implies that, apart from fundamental topics, considerable attention is paid to new plasma applications; plasma engineering in Europe is one of the main driving forces behind HTPP.

The conference supports the dissemination of methods for plasma diagnostics and monitoring and the exchange of models for plasmas sources and plasma applications. A novelty of HTPP 11 was the model market; a special type of poster session where running models were demonstrated and spectators were challenged to assemble their own plasma models using one of the available construction platforms.

For the first time in this series of conferences, the proceedings are published in two companion issues: *Journal of Physics D: Applied Physics*, which presents a selection of papers including invited and keynote papers, and the *Journal of Physics: Conference Series*. The present volume of the *Journal of Physics: Conference Series* includes 21 papers devoted to various branches of plasma physics. In line with the objectives of the HTPP conference, you will find papers on plasma sources, diagnostics and theory, covering the fields of thermal and non-thermal (even cold) plasmas, plasma–electrode interactions, surface treatment, synthesis, light generation and transport, and on applications in the fields of environmental technologies, biochemistry, and aeronautical and space sciences.

We would like to thank the members of the various committees, the participants who sent their contributions and the referees who did an excellent job giving support to improve the manuscripts.

We greatly appreciate the financial support from the conference sponsors: Association Arc Electrique, Belspo (Belgian Science Policy), Fonds National de la Recherche Scientifique, Ocean Optics Inc., Technifutur – Pôle Génie Mécanique & Solvay S.A..

Gérard Degrez	*Chairman of the Local Organizing Committee*
Joost van der Mullen	*Chairman of the Steering Committee*

Journal of Physics: Conference Series **275** (2011) 012001 doi:10.1088/1742-6596/275/1/012001

Characterisation of an Ar-H_2-O_2 ICP by OES: Measurement of the atomic concentrations of H and O

Altenberend J, Majchrzak M, Delannoy Y, Chichignoud G

SIMaP, BP 75, 1340 Rue de la piscine, 38402 Saint Martin d'Hères, France

J_altenberend@gmx.de

Abstract. In order to characterize an inductively coupled plasma torch used in refining of metallurgical silicon, we have developed a spectroscopic method based on absolute emissivity measurements and Abel inversion. This method permitted to measure the concentrations of atomic hydrogen and atomic oxygen, which are among the reactive species involved in the purification process. Assuming LTE, the temperature profiles are deduced from the emissivity of the Argon lines. The concentration of atomic oxygen is deduced from the intensity ratio O/Ar. The hydrogen concentration measurement has to take into account the Stark broadening and the Doppler broadening of the hydrogen lines. The comparison between experimental and simulated line profiles permits to determine this concentration. The method has been tested on 2 kW and 30 kW inductively coupled plasma torches at atmospheric pressure. The results show that the concentrations of atomic oxygen and atomic hydrogen can be measured with an accuracy of 25%. The main disadvantage of this method is that, using emission, it does not permit to measure the concentration in the "cold" zone of the plasma, i.e. at the edges.

1. Introduction

Today the major part of silicon for solar cells is purified by the Siemens process. Since this process is energy and cost intensive there is a lot of research is being performed with the goal of replacing it by cheaper and energy saving methods. Segregation works well for removing metallic impurities but boron segregates only very little. For this reason the laboratory SIMAP is working on a purification process for metallurgical silicon which reduces especially the boron concentration.

The process (see Figure 1) uses electromagnetic stirring to transport the impurities to the surface (i). There the impurities react with the reactive gases (ii) forming volatile reaction products, which are evacuated with the gas flow (iv). The plasma torch (iii) heats up the gases argon (~96%), hydrogen (~3%) and oxygen (~1%) to about 10000K and generates so the reactive gases. J Degoulange studied the reaction paths of the boron with the reactive gases [1].

He found that the radicals, which are likely to be important for the purification, are atomic hydrogen H, atomic oxygen O and the hydroxyl radical OH. He also found that the way of injecting the gases into the different flows changes the reactivity of the plasma on a carbon target [2]. By optimizing the temperature and concentration profiles one can so improve the efficiency of the process. Our laboratory is therefore developing a numerical model for the simulation of the plasma. In parallel we do measurements in order to improve the model but also to improve the process directly by comparing temperature and concentration profiles with the purification efficiency.

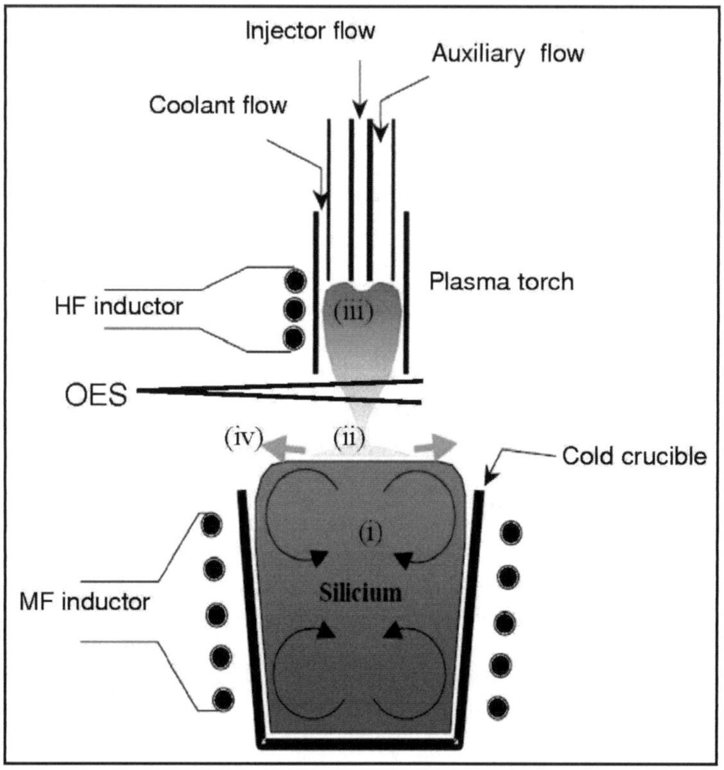

Figure 1 : The purification process :
(i) transport of the impurities to the surface
(ii) reactions between the reactive gases and the impurities
(iii) supply of the reactive gases from the plasma
(iv) evacuation of the reaction products

There are different ways to measure the temperature and the concentration which are compared in a paper of Wiese [3]. We apply a method based on atomic emission spectroscopy with Abel inversion for the measurement of the radial temperature profile and the radial concentration profiles of atomic oxygen and atomic hydrogen.

The spatially resolved concentration measurement by emission spectroscopy was yet applied in the 1970's to measure the spatial resolution of atoms in small plasma torches which are used for chemical analyses [4]. The interest of the research was focused more on the plasma parameters such as temperature, electron density and excited states density than on the concentration of the gases [5][6][7][8]. Therefore, and because of the assumption of LTE and symmetry, there are only a few papers using emission spectroscopy with Abel inversion for the measurement of the concentration. Other techniques like absorption spectroscopy and laser induced fluorescence (LIF) are often applied for the measurement of the concentration. The concentration of atomic oxygen and atomic hydrogen can be measured by two photons absorption laser induced fluorescence (TaLIF). This technique needs very high laser fluxes because the two photon absorption probability is weak and the spontaneous fluorescence which screens the TaLIF signal is high.

2. The method:

For the measurement of the temperature and the concentration we use the emissivity of atomic lines. We assume that the post-discharge zone of the plasma is in local thermal equilibrium (LTE). We will validate this assumption later with our results. In LTE one single temperature describes all energy modes (translational, electronic, ionization) for each coordinate. In plasmas this estimation is often false because of the high temperature gradients. Transport phenomena heat up and cool down the different energy modes differently. For example, electronic modes can be cooled down by emission of radiation for example. In atmospheric plasmas the collision rate is very high so that an LTE can be established between the different energy modes. For the establishment of LTE the transfer of energy by collisions must be much higher than the transfer of energy by electromagnetic radiation.

In LTE the Boltzmann distribution describes the distribution of the atoms on different electronic energy levels. The atoms in higher energy levels make spontaneous transitions to lower energy levels by emission of a photon. The emissivity of such a transition is the product of the concentration of the level, the transition probability A_{ul} and the energy of the photon $\frac{hc}{\lambda}$:

$$\varepsilon_{line} = n_u \cdot A_{ul} \frac{hc}{4\pi\lambda} = \frac{n_{tot} \cdot g_u \cdot exp(-E_u / kT)}{Z(T)} \cdot A_{ul} \frac{hc}{4\pi\lambda}$$

Thus, the emissivity depends so only on the concentration of the atoms and on the temperature. We know the concentration of argon which is very close to 100%. Therefore we measure the total emissivity of the argon line at 420.07 nm to calculate the temperature. Then we measure the emissivity of the oxygen line at 436.82nm and the hydrogen line at 486.13nm. We use the ratio between the emissivities of the oxygen line and the argon line for the calculation of the relative concentration of oxygen. We use the same method for the hydrogen concentration. For the choice of the lines we took care to avoid interference, self-absorption and low emissivity. The parameters of the lines can be found on the NIST website [9]. We use the Abel inversion [10] to calculate the emissivity profile from the lateral intensity profile of the plasma. The absolute intensity of light is only rarely used because of the difficulty of precise calibration. In our case this is not a problem because a high error of a factor two on the calibration gives only a low error of 4% on the temperature and 11% on the concentration as we calculated by multiplying the emissivity by two at a temperature of 9000K and an arbitrary concentration of atomic oxygen.

3. The experimental setup

3.1. The plasma torches

The two plasma torches each consist of three concentric tubes. The diameters are given in Table 1. The two outer tubes of the small torch a) are quartz tubes and the injector is made of alumina. The two inner tubes end at the height of the first turn of the inductor and the outer tube ends just above the last turn of the inductor and release the plasma into the air. We use a nebulizer to inject water into the small plasma torch and we can also inject oxygen. The inner and the outer tube of the big torch are water cooled copper tubes and the middle tube is a quartz tube. The outer tube is segmented in 8 parts so that it does not shield the magnetic field. It releases the plasma into a reactor chamber. The length of the injector can be adjusted to inject the gas directly into the plasma. We can inject argon, oxygen and hydrogen into the big torch.

Table 1 Geometry of the two plasma torches

	a)	b)
D1 external	30 mm	60 mm
D1 internal	27 mm	40 mm
D2 external	25 mm	37.5 mm
D2 internal	22 mm	33.5 mm
D3 external	4 mm	16 mm
D3 internal	2 mm	7 mm
D coil	37 mm	90 mm
L	41 mm	45 mm
Coil	5 turns	4 turns
P_{max}	2 kW	30 kW
Flow	12 l/min	40 l/min

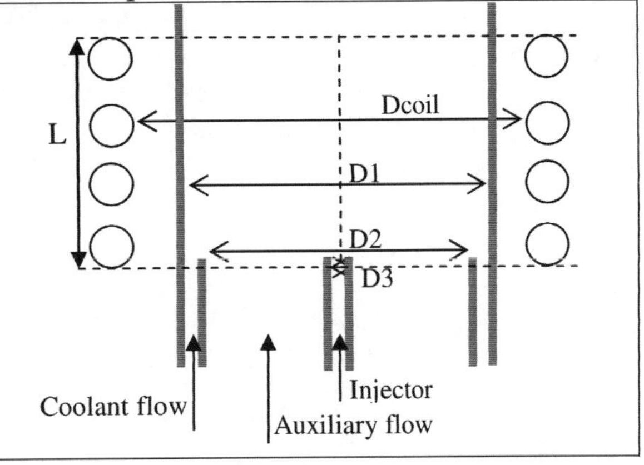

Journal of Physics: Conference Series **275** (2011) 012001 doi:10.1088/1742-6596/275/1/012001

3.1.1. The optical system

The light from the plasma is reflected by a mirror perpendicularly towards the monochromator. (See Figure 2) A lens focuses the light onto the entry slit of the monochromator. By turning the mirror we obtain a lateral intensity profile of the plasma. The distance between the mirror and the plasma is large compared to the diameter of the plasma. The small step size of the turning mirror allows us to measure get a lateral intensity profile of about 200 values for the small torch, which is good for the Abel inversion. For the calibration of the optical system we used a tungsten filament lamp after each measurement.

We opened the slits of the monochromator very wide (100μm) so that line broadening in the plasma is much smaller than the instrumental line broadening (FWHM=0.06nm). This works for the argon line and for the oxygen line. The Doppler broadening and the Stark broadening of the hydrogen line is large compared to the instrumental broadening. We used the assumption of LTE to calculate the width of the line with the measured temperature. For the Stark broadening we use the Saha equation for the electron density and a formula given by Griem [11] to calculate the line width. We assume that other broadening mechanisms are negligible.

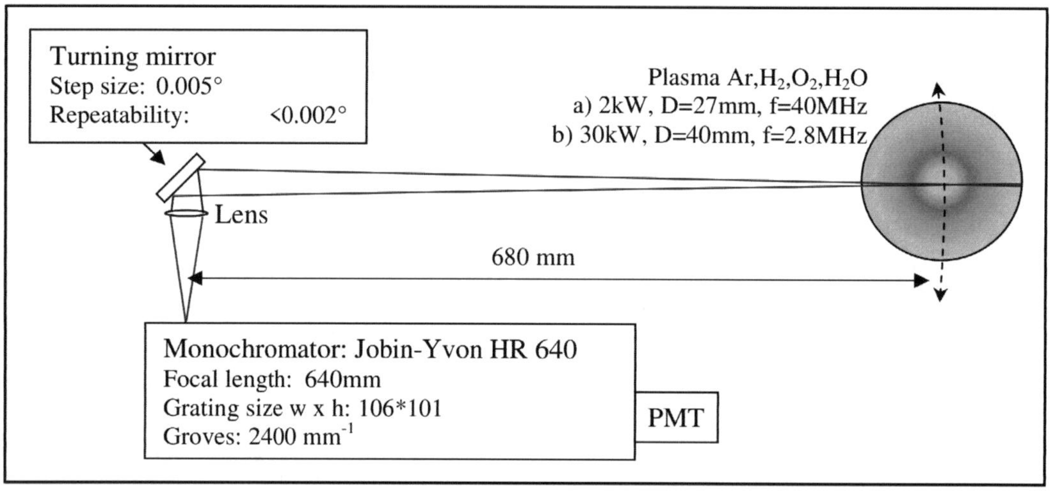

Figure 2 The optical system

4. Measurement of the line broadening:

We measured the line shape of the lines we used for our measurement ([11]). The argon lines and the oxygen line have exactly the same shape which is the shape of the instrumental broadening. The shape of the hydrogen line is the convolution of the Lorentzian profile from the Stark broadening, the Gaussian profile from the Doppler broadening and the profile of the instrumental broadening. For the measurement of the Stark broadening we calculated the emissivity at different wavelengths and fitted a convolution of the unknown Stark broadening with the known Doppler and instrumental broadening. In this way we obtain the full width at half maximum of the Stark broadening (Figure 4).

Journal of Physics: Conference Series **275** (2011) 012001 doi:10.1088/1742-6596/275/1/012001

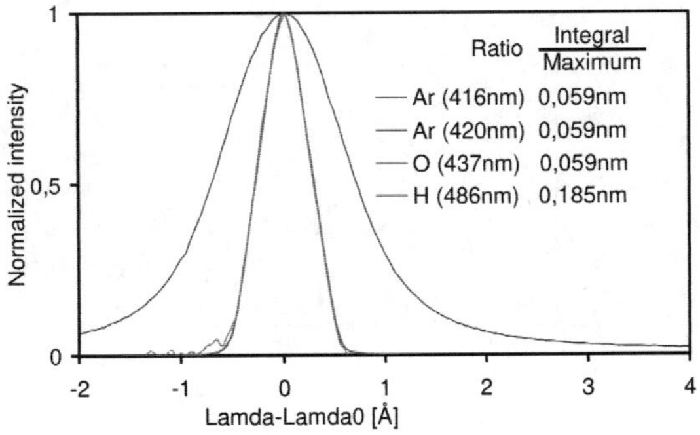

Figure 3 Shape of the lines

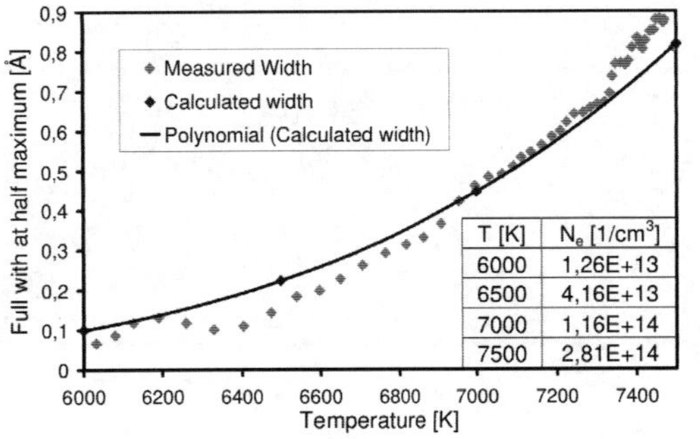

Figure 4 stark broadening of H

We can see that the measured line broadening is consistent with the calculated line broadening, especially at high temperature where line broadening is large. The small differences may be due to the measurement method. As we know exactly the line broadening of all the lines we can calculate the total emissivity of each line directly from the emissivity at the central wavelength of the line. We need this total emissivity for the concentration and temperature measurement.

5. Results

All measurements were made at the outlet of the torches. We began the measurements on the small plasma torch (2 kW). First we measured the temperature for two different argon lines with different transition probabilities (415.9nm and 420.1nm). If there was a radiative depopulation of these energy levels we could see a difference between the two temperature profiles (Figure 5). As there is no difference between the two temperature profiles the radiative depopulation of the energy levels is negligible. The error bars are due to the uncertainty in the transition probability which is better than 25%. For small temperatures the emissivity is very low and stray light makes the measurement impossible. Therefore our spectroscopic measurement is limited to a minimum temperature which is about 2000K below the maximum temperature.

We also injected water into the auxiliary flow in order to see if we can measure the concentration (Figure 6). The profiles are flat, which shows us that the gases are well diluted. The ratio of the two concentrations is close to two which shows us that the assumption of LTE is good. The increase of the concentrations at the edge of the plasma is probably due to the dilution with ambient air but also deviations from LTE may play a role. The error bars represent the difference between the right side and the left side of the lateral profile. It is due to temporal variations and to small deviations from the symmetry.

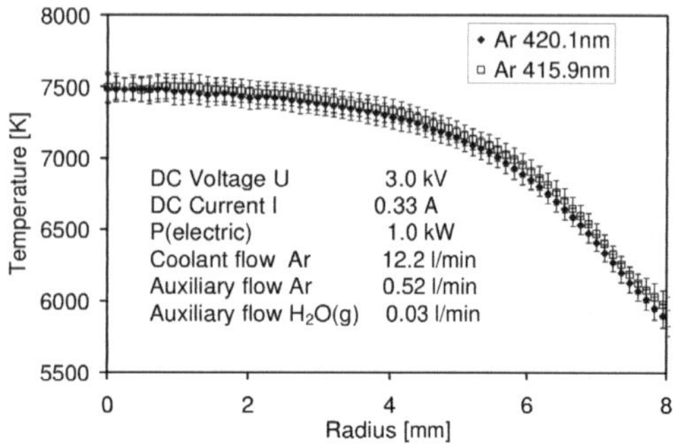

Figure 5 Temperature profile
(small torch, Ar +H_2O)

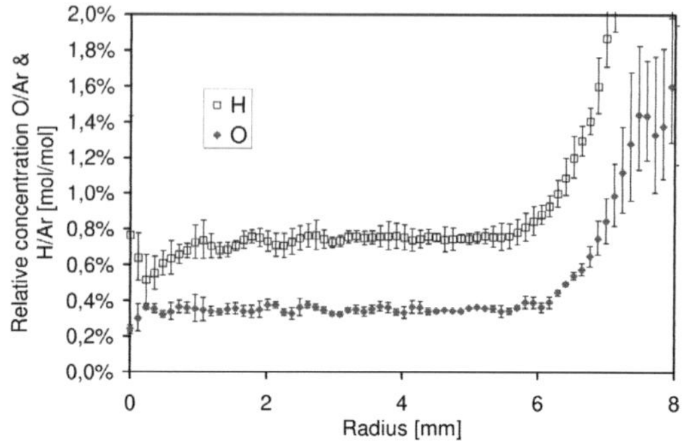

Figure 6 Concentration profile
(small torch, Ar +H_2O)

We also tried to measure a concentration gradient. For this purpose we injected a high flow of a mix of argon and oxygen in the injector. The effect is a strong cooling of the centre of the plasma and a bigger diameter of the plasma (Figure 7) and a concentration gradient (Figure 8). The measurement is not very precise in the centre of the plasma, because of the low emissivity and the high temporal variations.

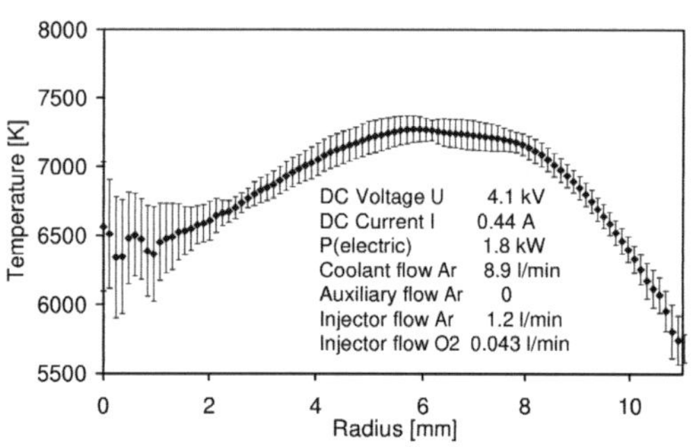

Figure 7 Temperature profile
(small torch, Ar +O_2)

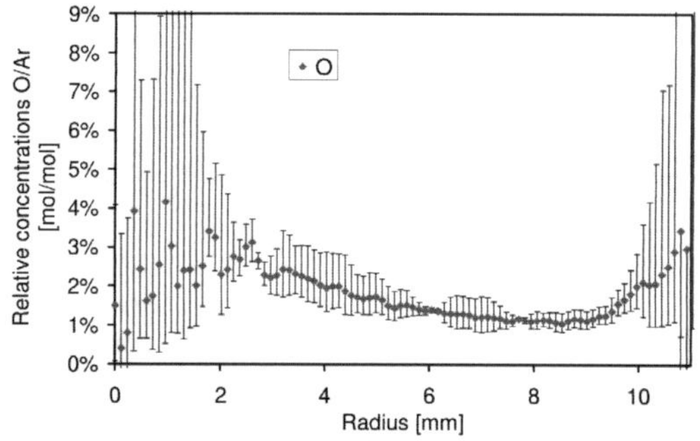

Figure 8 Concentration profile
(small torch, Ar +O_2)

We continued the measurement with the big torch. When we injected a relatively small oxygen flow in the injector we also saw a concentration gradient (Figure 10) but the cooling of the centre is less important (Figure 9). Since the concentration of oxygen in the measurement is very low the error bars are large. This is because of the high intensity of the background and of temporal variations. The measurement shows us that, with the injection of the reactive gases direct into the plasma, even a small flow rate results in a higher concentration in the centre. A higher concentration of reactive gases in the centre of the plasma is probably not good for the purification. We can use the method to study how we can get a better dilution of the gases.

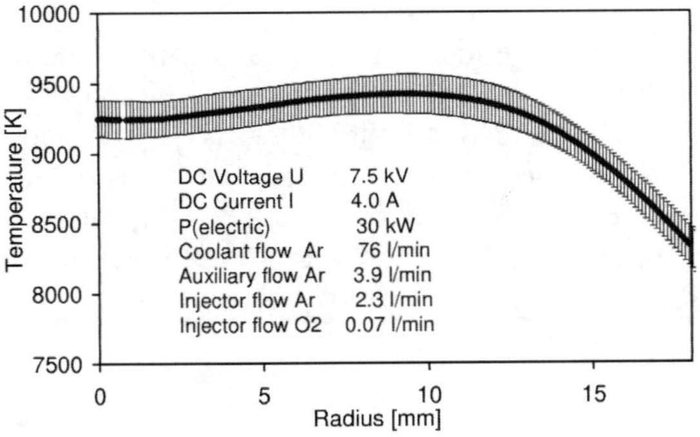

Figure 9 Temperature profile (big torch, Ar $+O_2$ in injector)

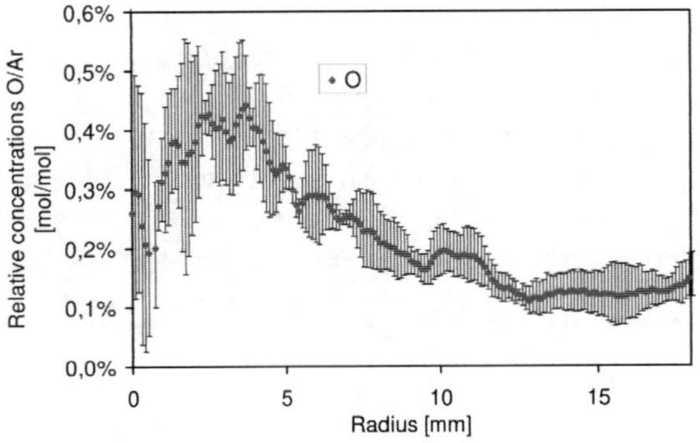

Figure 10 Concentration profile (big torch, Ar +O2 in injector)

In order to check the precision on the absolute concentration we injected a homogenous mix of argon and oxygen in all flows of the plasma torch so that we can compare the measured concentration to the injected concentration. We adapted the voltage of the generator so that the power injected rest constant. We show in Figure 11 the measured concentration for several different injected concentrations. The measured values are very close too the injected concentrations. This shows us that the plasma is very close to local thermal equilibrium and that our method works.

The differences can be due to different effects:

- the imprecision of the flow meters (rotameter)
- deviations from the symmetry
- temporal variations
- deviations from LTE

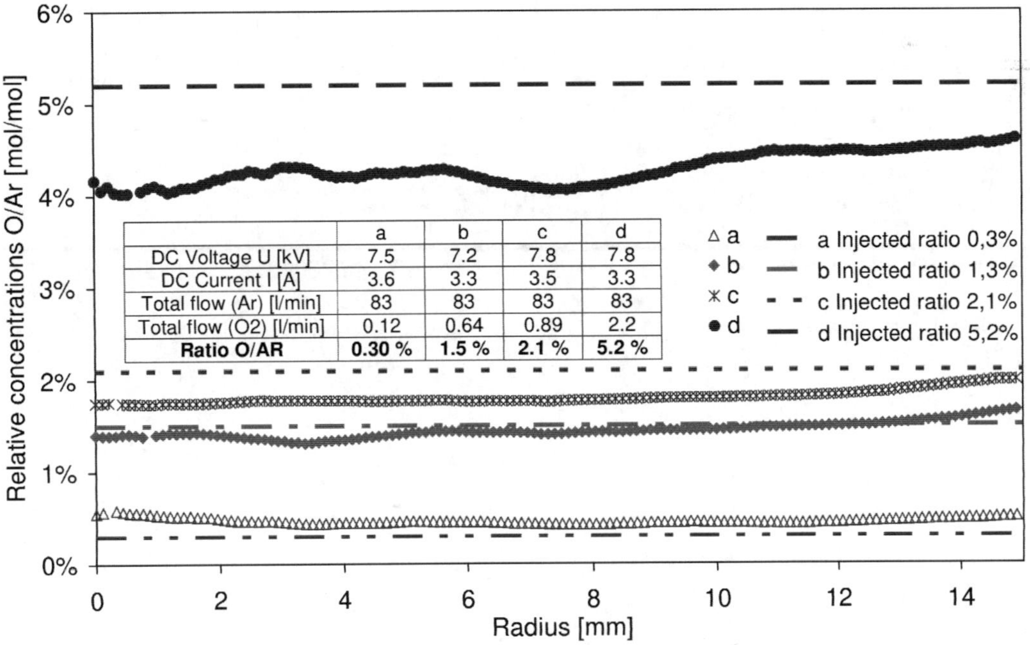

Figure 11 Concentration profiles of atomic oxygen for different injected ratios (Big torch, $Ar+O_2$)

6. Conclusion

The results showed that the examined plasma torches are very close too LTE. Therefore the method presented here permits the measurement of the concentration of the gases in an inductively coupled plasma torch with an absolute precision of better than 25%. The method is limited by stray light to the hot part of the plasma where the emissivity is high. The main sources of stray light are the backside of the reactor chamber and the mirror. These stray light sources can be reduced so that a study of the colder zones is possible. We showed that the gases in the process torch are not well diluted. Therefore it should be possible to use the method to optimize the dilution for the purification process.

References

[1] Degoulange J, 2008 *Purification et caractérisations physico-chimiques et électriques de silicium d'origine métallurgique destine à la conversion photovoltaique* Thèse de l'Institut National Polytechnique de Grenoble soutenue le 12 Décembre 2008

[2] Degoulange J et al, 2009 High *Temperature Material Processes*, 13(3-4): 315-324.

[3] Wiese W L, 1991 *Spectrochim. Acta;* 46B, 831-841

[4] Kornblum G R and Galan L, 1977 *Spectrochim. Acta;* 32B, 71-96

[5] Vacquie S, Gleizes A, Sabsabi M, Boulos MI, 1993 *Spectrochim. Acta;* 48B, 711-722

[6] Jonkers J et al 1999 *Phys. Scr.* 1999, 146

[7] Quintero MC, Rodero A, Garcia MC, Sola A, 1997*Applied Spectroscopy*, Vol 51, 778-784

[8] Lacombe JG, *Transfert radiatives dans les plasmas thermiques*, Thèse de l'Institut National Polytechnique de Grenoble soutenue le 10 Novembre 2008

[9] National Institute of Standards and Technology http://physics.nist.gov/PhysRefData/ASD/lines_form.html

[10] Nestor O H, Olsen H N, 1960, *SIAM Review*, Vol. 2, No. 3, pp. 200-207

[11] Griem H R, *Plasma Spectroscopy* New York: McGraw-Hill, 1964, 580p.

Journal of Physics: Conference Series **275** (2011) 012002 doi:10.1088/1742-6596/275/1/012002

Experimental quantification of the transient heat flux transferred to the electrodes in a carbon nanotubes synthesis reactor

Ramanantsoa R[1], Masquère M[1], Gonzalez JJ[1], Freton P[1] and Pacheco J[2]

[1] Laboratoire LAPLACE, UMR5213, UPS-CNRS-INP Toulouse III, 118 route de Narbonne Bât 3R2 31062 Toulouse cedex 9 (France)
[2] Instituto Nacional de Investigaciones Nucleares Carretera México-Toluca, CP 52750, México

E-mail: mathieu.masquere@laplace.univ-tlse.fr

Abstract. The aim of this work was to determine the energy exchange between the plasma and the electrodes in a carbon nanotubes synthesis reactor by applying an inverse method. Argon was used as plasma gas and the cylindrical electrodes were made of graphite. A space-marching method was used to recover the transient temperature profile in the electrodes and then to quantify the applied heat flux at their surfaces. Due to the high emissivity of graphite, lateral radiation of the electrodes was taken into account in the inverse model. The estimated heat flux deduced from the inverse method was validated by measuring the arc radiation losses obtained with radiant flux sensors. This approach provided power balance in the system. More than 60% of the power injected was transferred to the electrodes.

1. Introduction

Thermal plasmas created by an electrical arc are present in numerous industrial applications [1-3] and in laboratory experimental setups [4-10]. For many decades, experimental studies have been run to understand the phenomena occurring in arc plasma in order to optimize industrial processes such as cutting and welding systems, to enhance the quality of devices like circuit breakers and the resistance of different materials submitted to extreme conditions (aircraft under lightning, high-energy reactors, carbon nanotubes synthesis, etc. [e.g. 4, 11, 12]) and to validate numerical models. One point of interest concerns investigating heat transfer between the plasma and the electrodes. Carbon nanotubes synthesis by electrical arc is relatively recent; it is cheaper and more reliable than chemical approaches [12, 13]. However, the complexity due to numerous combinations of conditions and parameters [4, 11] implies that synthesis is still very difficult to control. Carbon nanotubes production is mainly influenced by anode erosion and by the plasma temperature gradients. An understanding of arc-electrode energy transfer provides a better interpretation of the results [4, 14]. This work deals with the experimental quantification of the heat flux deposited by thermal plasma on the electrodes in a carbon nanotubes reactor. The heat flux cannot be determined by direct measurement as sensors cannot resist

[1] Laboratoire LAPLACE, UMR5213, UPS-CNRS-INP Toulouse III, 118 route de Narbonne Bât 3R2 31062 Toulouse cedex 9 (France), mathieu.masquere@laplace.univ-tlse.fr

the high temperatures involved. Several numerical studies on arc power balance can be found in the literature [15-18], but only few deals with experimental measurements of the high heat flux deposited by the plasma on the electrodes. Direct measurement based on calorimeter devices [19] is one simple way to access the anode heat flux; this can also be used with an inverse method in more complicated systems [20]. In our configuration, due to lateral radiation losses, such approach was not used. Specific sensors like thermocouple-based devices [21, 22], sweeping wire devices, water-cooled probes [23], or calorimeters [24] could be exposed to the plasma flow but not for carbon nanotubes synthesis, it will modify the system configuration. We therefore decided to use an inverse method which consists of reconstituting the deposited heat flux profile by using measurement data taken from a more reachable zone of the electrodes. To this end, thermographic techniques and the data acquired were used in an inverse model for transient heat flux reconstitution. Similar investigations have involved a JET tokamak to control the flux deposited on tokamak's elements exposed to the plasma [25]. In that study, there was no need to consider material losses by radiation. Nevertheless, in our configuration, due to the nature of the electrodes and their high emissivity, the radiation on their lateral surfaces is taken into account in the model. In parallel with temperature history acquisition, the radiation power of the arc is recorded to be compared with the estimated heat flux. In the next sections, the experimental methods and the associated instrumentation will be presented before analyzing the results.

2. Description of the problem

The problem consists of measuring the transient heat flux deposited on the anode surface (lower electrode position) and on the cathode (upper position) by the thermal plasma. Due to the high flux level, direct measurement with a sensor cannot be performed and an inverse model is proposed (Figure 1).

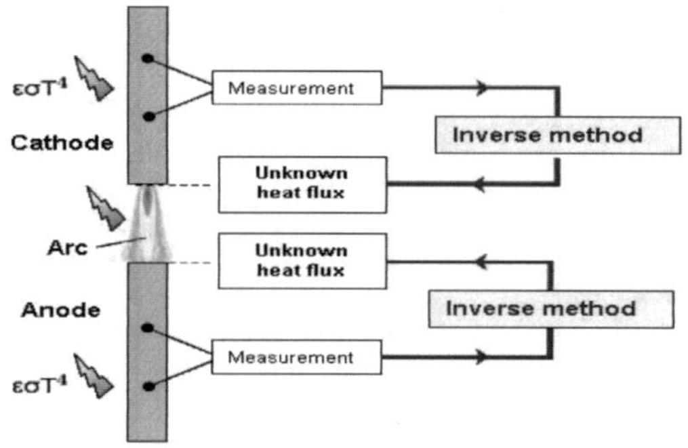

Figure 1. Illustration of the problem

For the carbon nanotubes reactor considered, the electrodes were made of graphite cylinders 6mm in diameter and about 7cm in length. The electrodes are assumed to be non-deformable so their lengths remain constant with time. Due to the electrode dimensions, combined with the fact that the thermal conduction inside the graphite electrode is mainly in the axial direction, the system can be assimilated to a one-dimensional problem. The electrode is thin enough to be able to ignore convection exchange with the ambient gas. However, graphite is a very emissive material and the thermal radiation from its surface should be taken into account.

3. The space-marching method

One-dimensional inverse problems have been the subject of many works using different kinds of methods and various approaches [12-19, 24, 31-33]. We used the method described by Raynaud [30] because of its non-iterative approach. It is a variant of the space-marching method established by D'Souza [32] with a Crank-Nicolson implicit scheme. To reconstitute the transient profile of the applied heat flux, the temperature histories in two points of the electrode must be known. Therefore the method divides the calculation domain into two zones: the direct zone, delimited by the two sensors, and the inverse zone between the first sensor and the active surface. First, the direct zone was dealt with and the temperature map versus time was deduced from the measured temperatures. Therefore, the inverse method was solved by scanning the inverse zone step by step starting from the first sensor location and moving towards the surface as shown in figure 2.

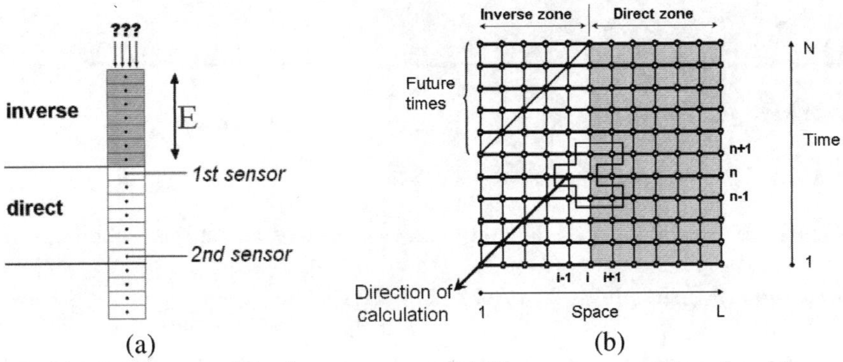

Figure 2. **(a)** The direct and the inverse zone. **(b)** The space marching algorithm

The method proposed is simple and straightforward to implement since the characteristics of the material are known. Both the direct and the inverse zones are solved starting from the energy balance in each cell at each time. The smaller the time step Δt, the greater the accuracy of the method and the faster the temporal variations that can be reconstituted. However this time resolution is limited by the dimensionless time step $\alpha\frac{\Delta t}{E^2}$ where α is the diffusivity of the material [$m^2.s^{-1}$], Δt the time step [s] and E [m] the distance from the electrode surface to the first captor position. This expression is indirectly correlated to the Fourier number $M=\alpha\frac{\Delta t}{\Delta x^2}$ (Δx the space step size).

So, for a given problem, the space and time resolution should be considered to obtain a compromise between the stability, the precision and possibly the computing time. Numerical tests and parametric studies were investigated in a previous study to find the best combination of parameters for a good experimental result [34]. For the inverse method, the thermophysical properties of the material must also be known. The heat capacity and the thermal conductivity that we used come from the Touloukian, the Janaf and the NIST databases [35-37].

4. Experimental setup

Measurements were made in a fullerene and carbon nanotubes synthesis reactor [4]. The gap between the 6mm-diameter graphite cathode and anode was roughly 5 to 6mm. The plasma medium was composed of argon at atmospheric pressure. Arc voltage and current were measured to calculate the total power injected into the system.

The temperature measurements were made with a FLIR SC6000 infrared camera. In figure 3 (left side) we can observe the experimental setup consisting of the reactor and the camera. The camera lens was located 40cm from the electrode and had space resolution of about 0.2mm for this configuration. The two points of measurement should have temperatures in the range in which the camera operates (600°C-1200°C).

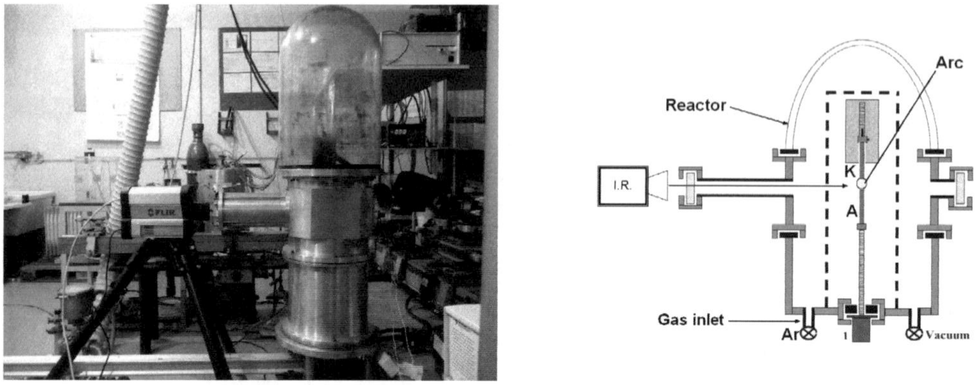

(a) (b)

Figure 3. (a) The IR camera and the reactor **(b)** Cross-sectional schema of the reactor

Estimation of the spectral emissivity of the electrodes

To obtain accurate temperatures, the emissivity of the object needs to be known in the interval 3µm-5µm. So, the camera emissivity entry is adjusted to correspond to the temperature given by one thermocouple. This approach allows us to estimate the spectral emissivity of the electrode at 0.85 for the spectral range of the camera. This value is very close to that given by other authors [38, 39].

Radiative power measurement

As convection in the system is assumed to be negligible, the total power injected into the arc should be the sum of the power transferred to the electrodes and the power lost by the arc through radiation. $P_{arc} = V_{arc}.I_{arc} = P_{electrodes} + P_{rad}$. This expression was used to cross-check the estimated heat flux. We therefore used radiant flux sensors with a large spectral width (0.1µm-12µm). Due to the anisotropy of the arc radiation, the sensors were placed along the azimuthal direction (figure 4). The high emissivity of the graphite in the wavelength range of the captors, implies they receive the flux not only from the arc but also from the electrodes.

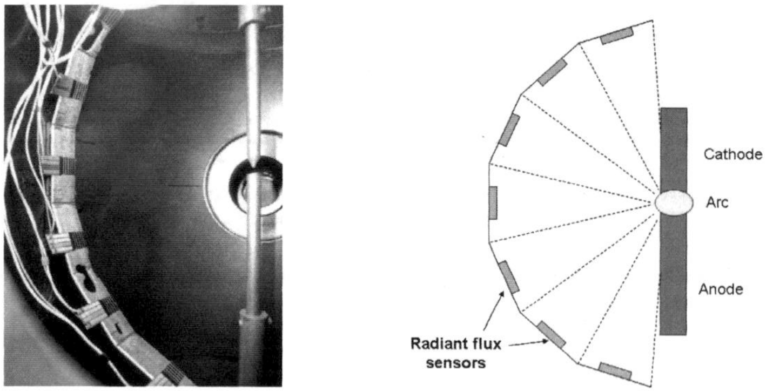

Figure 4. Assembly of radiant flux sensors

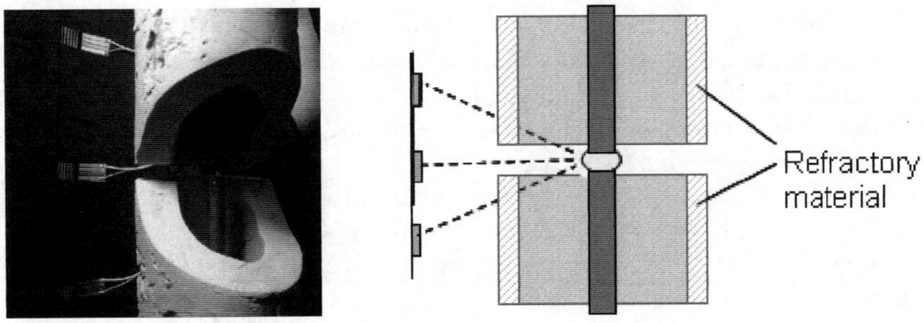

Figure 5. Terracotta mask for hiding the electrodes

The percentage radiation emitted by the electrodes was obtained by masking them with a refractory material (clay) as shown in Figure 5. The masks were placed far enough away to keep them cool and to avoid fusion. Two other sensors were placed above and below to check that there was no flux produced by any reflection phenomena on the reactor wall. If we note F and F' the flux received by the central captor respectively without and with a mask, the ratio F/F' represents the percentage of arc radiation in the total measured flux. This ratio is equal to 40% and is assumed to be the same for each sensor.

5. Heat flux quantification

In this section we report the results of applying the inverse method to experimental data. The temperatures were picked up 20mm and 25mm from the anode surface, and 10mm and 15mm from the cathode tip (Figure 6). We worked with a DC power supply with current varying from 20A to 50A in 10A steps. The current intensity range was chosen so as to avoid electrode erosion. Indeed this physical mechanism, which could change the electrode length, is not taken into account in our inverse method.
Multiple temperature channels are proposed with the camera. We selected the range [600°C-1200°C] as it simultaneously provides four appropriate measurement points for the cathode and the anode. Two cathode configurations were tested: one where the end of the cathode was flat and the other where cathode was pointed. The pointed cathode tip provided a more stable arc but the results for both setups were practically the same.

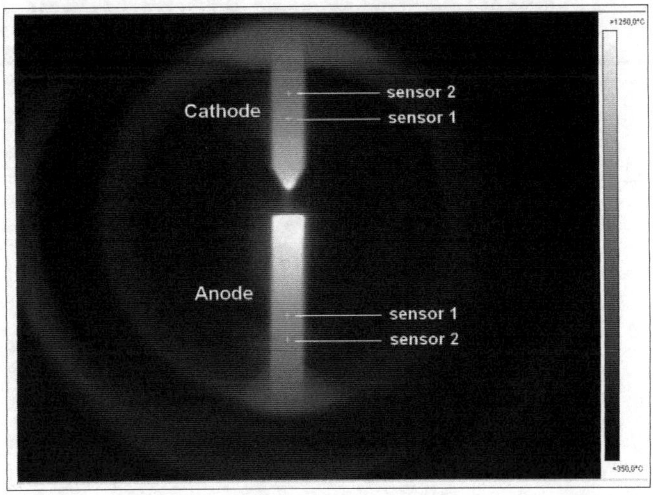

Figure 6. Image from the infrared camera

Below (Figure 7) we report the results obtained introducing experimental data into the inverse model. Figure 7a gives the temperature histories by thermography. The power transferred by the plasma to the electrodes is plotted in figure 7b. It is deduced from the integration of the calculated heat flux over the effective section of the electrodes. The uncertainty affecting the results is mainly due to the position of the measurement points for temperature assessment (±0.2mm) and to the spectral emissivity of the electrodes (between 0.83 and 0.87). To determine the uncertainty, the method was applied with the two extreme combinations. The ratio between the power deposited in the electrodes and the total power injected was between 60% and 80% (Figure 7c). The incident power to the cathode represented only a small proportion and was only weakly dependent on the power injected. Most of the power deposited was focused on the anode. Figure 7d shows that the variation of the anode power versus the arc current was practically linear, as reported in other experimental works [10, 19, 40].

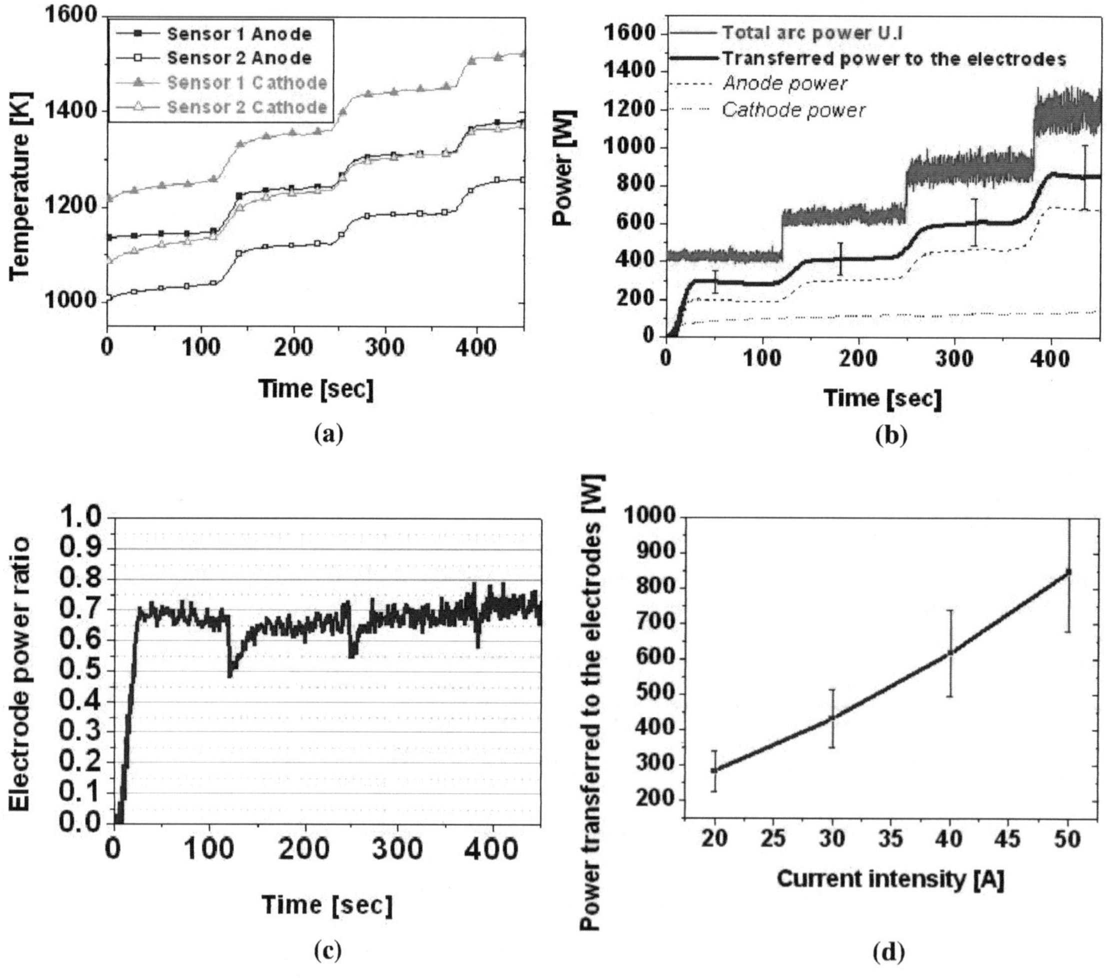

Figure 7. (a) Temperature histories **(b)** Power transferred to the electrodes **(c)** Ratio of the power transferred to the electrodes from the applied power **(d)** Power transferred versus arc current.

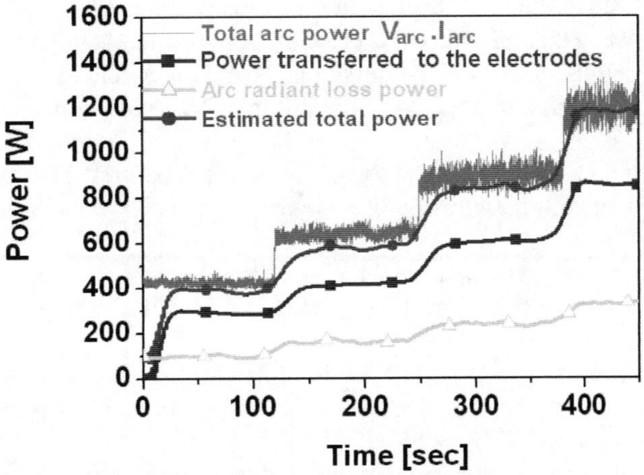

Figure 8. Verification with radiant flux measurement

To check our results, we plotted (figure 8) the total power injected into the system, the total power estimated from the inverse method giving the power transferred to the electrodes and the radiation power. The calculated total power was slightly underestimated. This could be because we chose to ignore convection.

6. Conclusions

Experimental measurements were applied to quantify heat transfer between plasma and electrodes in a carbon nanotubes reactor. A space-marching inverse model was associated to thermographic acquisition to assess the incident heat flux on the surface of the electrodes. In parallel, arc radiation losses were measured to check the results. This allowed us to calculate the power balance in the configuration considered. Good agreement was found between the radiation losses measured directly and those calculated from injected power minus the power transferred to the electrodes. The power transferred to the electrodes was estimated at about 60% of the injected power. This approach can be extended to other problems since the material properties are known. Before being applied as a metrology tool for carbon nanotubes monitoring, the method needs to be improved by taking the erosion into account. The inverse method should be suitable, and the next step of the study will focus on erosion.

7. Acknowledgement

The authors would like to thank the Ecos-Nord program for supporting this work under project No. M08-P01.

References

[1] Nadkarni S V 2008 *Modern Arc Welding Technology* (Ador Welding Ltd)

[2] Harris I D 1997 *Plasma arc cutting of bridge steels* (Transportation Research Board)

[3] Laroche G 1991 *Les plasmas dans l'industrie* (Dopee Diffusion, Electricité de France)

[4] Pacheco M 2003 *Synthèse des Nanotubes de Carbon par Arc Electrique*, PhD Thesis, Université Paul Sabatier, Toulouse, France

[5] Fauchais P 2004 Understanding plasma spraying *J. Phys. D: Appl. Phys.* **37** R86

[6] Teulet P, Girard L, Razafinimanana M, Gleizes A, Bertrand P, Camy-Peyret, Baillot E and Richard F 2005 Experimental study of an oxygen plasma cutting torch: II. Arc–material interaction, energy transfer and anode attachment *J. Phys. D: Appl. Phys.* **39** 1557

[7] Tanaka M and Ushio M 1998 Plasma state in free-burning argon arc and its effect on anode heat transfer J. *Phys. D: Appl. Phys.* **32** 1153–1162
[8] Zielinska S, Pellerin S, Valensi F, Dzierzega K, Musiol K, De Izarra Ch, Briand F 2008 Gas influence on the arc shape in MIG-MAG welding *EPJ. Applied* physics vol. **43**, no1, pp. 111-122
[9] Wilden, J., Bergmann, J.P., Frank, H 2006 Plasma Transferred Arc Welding—Modeling and Experimental Optimization *Journal of Thermal Spray Technology*, Vol **15**, Number 4 pp. 779-784
[10] Gonzalez JJ, Freton P and Masquère M 2007 Experimental quantification in thermal plasma medium of the heat flux transferred to an anode material J. Phys. D: Appl. Phys. 40 5602–5611
[11] Mansour A, Razafinimanana M and Monthioux M 2009 Influence of the anode filling carbon granulometry on plasma features in the electric arc process applied to carbon nanotubes *synthesis J. Phys. D: Appl. Phys.* **42** 195203
[12] Ostrogorsky AG, Marin C 2006 Heat transfer during production of carbon nanotubes by the electric-arc process *Heat Mass Transfer* **42** 470–477
[13] Du F, Ma Y, Lv X, Huang Y, Li F, Chen Y 2006 The synthesis of single-walled carbon nanotubes with controlled length and bundle size using the electric arc method *Carbon* **44** 1327-1330
[14] M Keidar1, A Shashurin1, O Volotskova1, Y Raitses and I I Beilis 2010 Mechanism of carbon nanostructure synthesis in arc plasma *Phys. Plasmas* **17** 057101
[15] Amakaway T, Jenista J, Heberleinx J and Pfender E 1998 Anode-boundary-layer behaviour in a transferred, high-intensity arc *J. Phys. D: Appl. Phys.* **31** 2826–2834
[16] Teste P, Leblanc T, Rossignol J, Andlauer R 2008 Contribution to the assessment of the power balance at the electrodes of an electric arc in air *Plasma Sources Sci. Technol.* **17** 035001
[17] Lago F, Gonzalez J J, Freton P, Ihlig F, Lucius N and Piau G P 2006 A numerical modelling of an electric arc and its interaction with the anode: III. Application to an interaction of a lightning strike and an aircraft in flight *J. Phys. D: Appl. Phys.* **39** 2294–2310
[18] Perez R, Rojas H, Walder G, Flükiger R 2004 Theoretical modeling of energy balance in electroerosion *Journal of Materials Processing Technology* **149** 198–203
[19] Sanders N A, Pfender E 1984 Measurement of anode falls and anode heat transfer in atmospheric pressure high intensity arcs *J. Appl. Phys.* **55** (3) 714-722
[20] Carpentier S, Le Niliot C, Chantant M, Corre Y, Gardarein JL, Rigollet F 2007 Calcul inverse de puissance moyenne incidente sur un composant activement refroidi à partir de données calorimétriques *Symp. SFT* Île des Embiez France
[21] Kanzawa A 1975 Distributions of current density and heat flux around a cylindrical probe in an atmospheric pressure plasma *Heat Transfer - Japanese Research, vol.* **4** 37-44
[22] Cheron BG, Robin L and Vervisch P 1991 Parietal heat flux measurement in a nitrogen plasma jet *Meas. Sci. Technol.* **3** 58-61
[23] Pfender E 1976 Heat transfer from thermal plasmas to neighboring walls or electrodes *Pure & Appi. Chem., Vol.* **48** 199-213
[24] Löhle S, Battaglia JL and Gardarein JL 2008 Heat flux measurement in a high enthalpy plasma flow *Journal of Physics: Conference Series* **135** 012064
[25] Corre Y, Andrew P, Gardarein JL, Hellner T, Le Niliot C, Poyet M, Rigollet F 2007 calcul inverse du flux de chaleur dans le tokamak JET par méthode de retour vers la surface : approche 1D non linéaire *Symp. SFT* Île des Embiez France
[26] Ahmadizadeh Y, Soti V, Pourgholi R 2007 Numerical solution of an inverse diffusion problem *Applied Mathematical Sciences,* Vol. **1**, 2007, no. 18, 863 - 868
[27] Molhem H and Pourgholi R 2008 A Numerical algorithm for solving a one-dimensional inverse heat conduction problem *Journal of Mathematics and Statistics* **4** (1) 60-63

[28] Burggraf O.R 1964 An exact solution of the inverse problem in heat conduction theory and applications *Journal of heat transfer* **86C** 373-382

[29] Taler J and Duda P 2001 Solution of non-linear inverse heat conduction problems using the method of lines *Heat and Mass Transfer* **37** 147-155

[30] Raynaud M and Bransier J 1985 A New Finite-Difference Method for the Nonlinear Inverse Heat Conduction Problem, *Numerical Heat Transfer Vol* **9** 27-42

[31] Carasso A.S Space marching difference schemes in the nonlinear inverse heat conduction problem *Inverse Problems* **8** 25-43

[32] D'Souza N 1975 Numerical solution of one-dimensional inverse transient heat conduction by finite difference method, *ASME paper* **75**-WA/HT-81

[33] Hensel E. C, Jr., Hills, R. G 1984 A space marching finite difference algorithm for the one dimensional inverse conduction heat transfer problem *Conf. ASME, National Heat Transfer Conference, Niagara Falls, NY, Aug.* 5-8

[34] Ramanantsoa R, Masquère M, Gonzalez JJ, Freton P, Franceries X 2009 Quantification transitoire du flux d'énergie déposé par un plasma thermique *Symp. CAE* IX MINES ParisTech, Sophia Antipolis

[35] Touloukian, Y. S.1970 Thermal conductivity: nonmetallic solids *Thermophysical properties of matter,* **vol 2** 20-52

[36] Chase M.W, Jr.,. Davies C.A,. Downey J.R, Jr,. Frurip D.J,. McDonald R.A, and. Syverud A.N 1986 JANAF Thermochemical Tables Third Edition, Part I, 535-536

[37] Hust JG and Lankford AB 1984 Thermal conductivity and electrical resistivity as a tuncion of temperature (IPTS-68 and NBS P2-20) from 5 to 2500 K, Nat. Bur. Stand., Report of investigation, Research Materials 8424, 8425 and 8426 Graphite

[38] Savvatimskiy A.I 2003 Measurements of the melting point of graphite and the properties of liquid carbon (a review for 1963–2003) *Carbon* **43** 1115–1142

[39] Neuer G and Jaroma-Weiland G 1998 Spectral and total emissivity of high-temperature materials *International Journal of Thermophysics* Vol. **19** 917-929

[40] Bini R, Monno M and Boulos MI 2006 Numerical and experimental study of transferred arcs in argon *J. Phys. D: Appl. Phys.* **39** 3253–3266

Journal of Physics: Conference Series **275** (2011) 012003 doi:10.1088/1742-6596/275/1/012003

Last developments in diagnostics to follow splats formation during plasma spraying

S. Goutier, M. Vardelle, P. Fauchais

SPCTS Laboratory, University of Limoges, 123 Avenue Albert Thomas, 87060 Limoges, France

E-mail: simon.goutier@unilim.fr

Abstract. Plasma sprayed coatings adhesion-cohesion, as well as thermo-mechanical properties depend strongly upon the sprayed particle flattening and solidification , taking place in a few μs and the corresponding splat formation. Sophisticated fast pyrometers (response times of less than 100ns) have been developed for studying the flattening particle temperature evolution when impacting on smooth surfaces. However the signal interpretation is far to be easy without following simultaneously the evolution of the particle shape. As the flattening time is in the μs range and the acquisition frequency of the fastest camera is only one image/μs, it is not possible to follow the flattening particle shape evolution. That is why, different particles considered to have the same impact parameters, were photographed at different flattening times separated by a few hundreds of nanoseconds. In parallel measurements on millimetre-sized drops (metals and ceramics), which flattening time is in the ms range, have been developed because fast cameras (~ 5000 images/s) can be used to follow the flattening phenomena. In this paper are presented the experimental devices developed and impacts of alumina particles in the 40μm and 5mm ranges, but with Weber numbers as close as possible, are compared. Measurements at both scales demonstrate the importance of the substrate preheating control (time and temperature) to achieve disk shaped splats and explain the phenomena induced.

1. Introduction

In thermal spray the coatings are build up by successive impacts of particle, which form splats after flattening and solidification. So coating properties depend mainly of the adhesion between splats and between splat and substrate. The link between the first splats layer and the substrate plays the main role on the coating adhesion. To improve the contact between splat and substrate, some authors studied the splat obtained and particularly its shape. If this technique gives an idea of some trends, it is difficult with these results to conclude on the influence on splat generation of the different in flight parameters. Indeed when the substrate surface temperature is increased over a given temperature, called transition temperature, splats are disk shaped. In fact it exists a temperature for which the proportion of circular and fragmented particles is identical. However it is mandatory to associate the in-flight particle parameters to splats formation to draw sound conclusions on the substrate temperature effects, which is not yet really possible. So to better understand the involved phenomena, many works have been devoted to the study of a single splat formation; see the reviews of Fauchais and al [1, 2]. They have emphasized the difficulty of the study: impact of particles in a molten state with sizes between 20 and 60 μm with velocities ranging between 50 and 600 m/s, resulting in flattening and solidification times in the μs time range. If sophisticated pyrometers have been developed with response times of about 100ns, the interpretation of their signals is far to be straight forward without the visualization of the flattening phenomenon. In spite of the development of fast cameras, it is still impossible to follow this flattening for the same micrometer sized particle. That is why, since the mid nineties, many works have been devoted to the flattening of millimeter sized metal drops [1, 2] occurring in the millisecond time range instead of the microsecond one. However most results were obtained with free falling drops (velocities below 3m/s) resulting in Weber numbers much lower than those obtained in plasma spraying conditions. The comparison of phenomena at both scales (mm and μm sized) is thus rather difficult. If for tin drops, using a centrifugal droplet set-up, impact velocities up to 30 m/s have been achieved [3], this material is unfortunately not sprayed by thermal processes.

The main difficulty is to achieve imaging at the micrometer scale. To our knowledge only one author has used a very fast camera (1 frames/μs) for following the flattening of a single zirconia particle impacting on a cold substrate [4]. Some authors have also studied the impact of metal and ceramic particles on glass or inconel substrates at different times of the flattening process but for different particles [5-6].

In this paper are presented first the experimental devices developed and the comparison of the impacts of alumina particles in the 40 μm and 5 mm ranges, but with Weber numbers as close as possible. Measurements at both scales demonstrate the importance of the substrate preheating control (time and temperature) to achieve disk shaped splats and explain the phenomena induced. At last some examples of the different impacts are presented.

2. Experimental setup

Experimentations are carried out with two experimental techniques for studying impact and flattening phenomena:

-The first one is a modified free-falling set-up to study millimeter-sized drops with low or high Weber numbers.
-The second one uses a plasma spraying set-up with micrometer-sized drops.
Running in parallel these two studies enable comparing particle flattening and cooling on smooth 304L substrates, preheated or not, at spatial and time scales differing by almost three orders of magnitude.

2.1. Millimeter system

To produce liquid ceramic drops, an alumina rod is introduced and melted in an electrical arc furnace (fig 1). A suspended drop is formed. When the gravity force overcomes that of the surface tension, the drop falls. This set-up is disposed into an argon controlled atmosphere chamber.
Thanks to the use of a jack, the substrate can be moved up and the relative impact velocity can reach up to 10 m/s [7].

A detector located at the chamber output, generates a TTL pulse when one single drop crosses its measuring volume. This pulse is then directed to the measuring system composed of a fast camera (Photron 4000 image/s) targeting the substrate in two configurations, first tangentially in order to follow the matter ejection during the drop flattening and second with an angle of 45° to follow the whole flattening. Of course, for the first position, in order to better determine the location of ejections, the camera isn't absolutely parallel to the substrate but a small angle (2°) exists, allowing seeing the opposite edge of the splat.

Finally, during the movement of the substrate, a fast "on board" two-color pyrometer fixed on a jack controls the drop temperature at impact. The main difficulty is due to the abrupt deceleration of the substrate, which permits to study the flattening only during the first 14 ms.

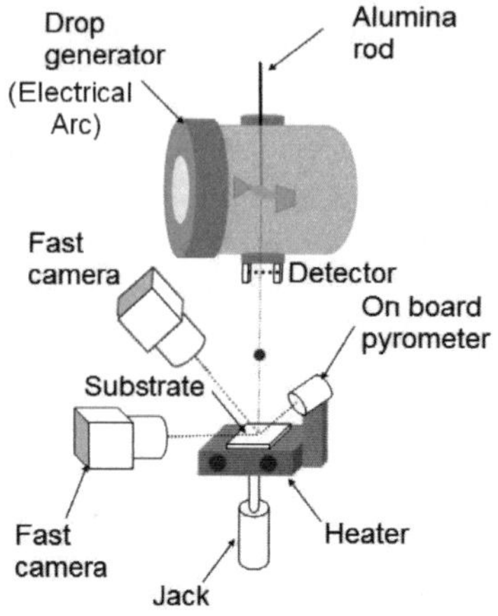

Figure 1. Drop generator for millimetre sized particle

2.2. Micrometer system

Particles are plasma sprayed using a direct current (D.C.) plasma torch (PTF4 type) with a 6 mm internal diameter anode-nozzle and running with a mixture of argon-hydrogen. The arc current is 650A, the argon flow rate 33 L/min and the hydrogen volume percentage 25%. An alumina powder with particle sizes between 40 and 50 μm is used in this study. These spray conditions result in fully melted particles with velocities at impact around 200 m/s. Splats are collected on a smooth (Ra=0.06 μm) 304L substrate fixed 110 mm downstream of the nozzle exit (figure 2).

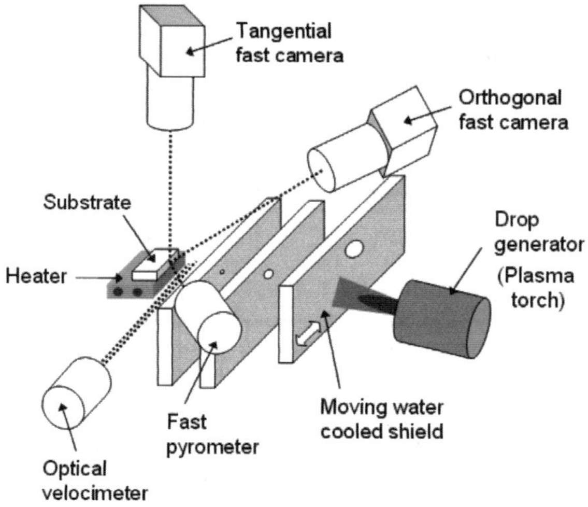

Figure 2. Drop generator for micrometer sized particle

Three shields are used to select particle trajectories close to the plasma axis:
- A moving water cooled shield (hole of 5 mm) fixed 70 mm downstream of the torch nozzle exit
- Another one (hole of 1.5mm) fixed 10 mm behind the first shield.
- A last one with a hole of 600 μm fixed 10 mm upstream of the substrate.

2.2.1. In flight velocity measurement system

The in-flight velocity characterization is made by a two points measuring optical detector, the signal treatment permitting to filter the raw signal coming from PM, and generating a TTL pulse on the falling edge of each smooth signal (see figure 3). These two signals, thanks to specific hardware, called velocimeter and running under LabView, permit to measure the time between both TTL pulses. After the velocity calculation, the hardware calculates at which time the particle will impact the substrate. Finally a pulse to start the different measurement systems at a given time (mainly at the impact) is generated. The relative random error on velocity is about 10 % and 5% on the acquisition triggering.

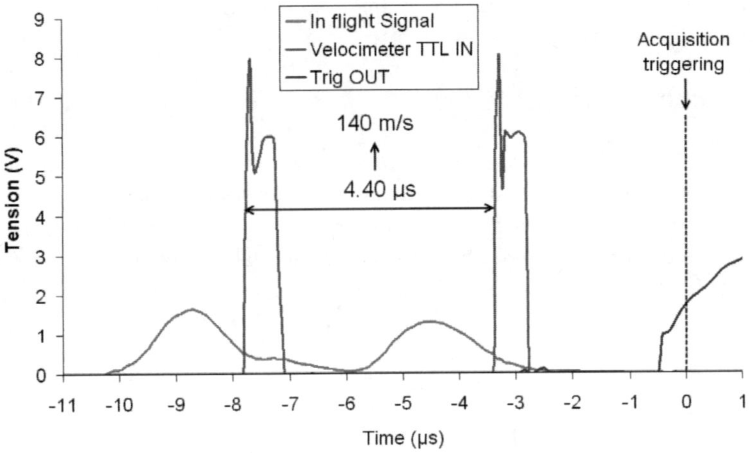

Figure 3. Example of in-flight velocity measurement for an alumina particle 40 μm in diameter with an impact velocity of 140 m/s. The triggering acquisition is the reference time for all measurement systems (pyrometer and cameras)

2.2.2. Particle temperature in flight and during the flattening

A fast (50ns) pyrometer follows the particle temperature during its flattening and cooling. The two wavelengths of the pyrometer are 690 μm and 710 μm. A smoothing is applied on the two original signals (for two wavelengths), but in order to avoid a modification of the profile curves, there are three smoothing areas: - the first area, noted 1 in figure 4, corresponds to the particle signal during its flight just before the impact, - the second area, noted 2, corresponds to the particle flattening on the substrate, - finally, the last area, noted 3, corresponds to the splat cooling and / or the splat diameter reduction. After that, the hardware calculates the dichromatic temperature. When the signal begins to decrease the temperature is calculated thanks to monochromatic function.

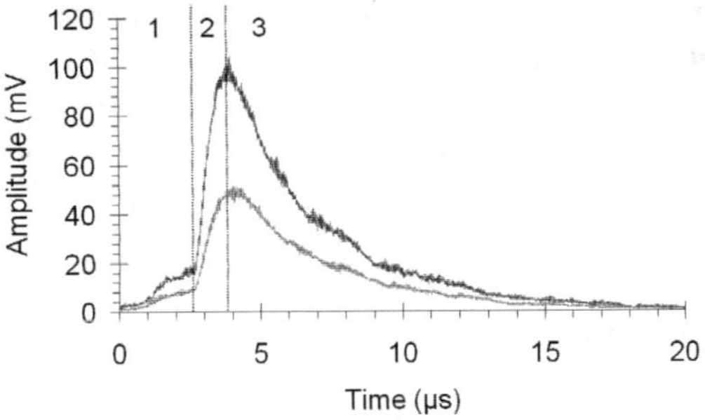

Figure 4. Position of the different smoothing area on a typical thermal signal coming from pyrometer

Journal of Physics: Conference Series **275** (2011) 012003 doi:10.1088/1742-6596/275/1/012003

2.2.3. Particle flattening by imaging technique

Imaging techniques allow following the particle flattening. On each camera are fixed a macro lens with a focal distance of 180 mm and two converter (X2) lenses. The main advantage of these lenses compared to a long distance microscope (type Questar) is the more important collected light. One camera is disposed orthogonally to the substrate and other one tangentially to it. The main problem is due to the movements of high temperature gas produced by the plasma torch, which modify the refractive index and thus the focal point of both cameras. It is then necessary to use some cooling air system disposed in front of the camera in order to maintain a constant temperature of the optical path to keep a refractive index as constant as possible.

By adding a variable time delay to the calculated impact time, it is then possible to observe the flattening of micrometer sized particles at different times but for different particles (of course assuming they have the same impact parameters).

2.2.4. Heating and cooling of substrate

A monochromatic pyrometer (Ircon 5 μm, 10 ms of response time) controls the substrate temperature during the preheating stage, an automatic system permitting to keep a given temperature.
In order to study the effect of the substrate oxidation it is necessary to freeze its oxidation state. For that the heating system is shifted to a cooling one, which permits to come back to room temperature in less than 2 minutes (see figure 5). If the cooling system is not used the substrate comes back to room temperature after 20 minutes.

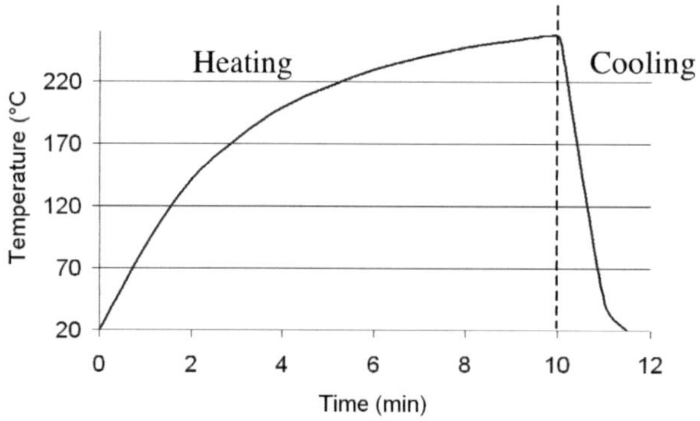

Figure 5. Typical signal of heating and cooling for a stainless steel substrate. $T_{set\ point}$ =250°C $t_{preheating}$ = 10 mn

2.2.5. Typical procedure to study micrometer sized particle impact.

With this system, the total time of an experiment, has to be very short. Indeed when the plasma torch is running too long, all optical adjustments are modified due to the hot temperatures, which induce the expansion of the metallic parts. So the spray system is controlled by a homemade hardware, following and controlling all successive steps to obtain one single splat. The main step consists in closing very fast (about 5 ms), the powder feeder and the mobile shield when one particle crosses the velocity measuring area, to avoid several particles on the measurement area. The total time to obtain a single splat onto the substrate is about 10 s.

2.3. Substrate

For both particle sizes, substrates are made of stainless steel (304L) and mirror polished by using SIC paper 4000. They are fixed on copper supports that can be heated up to 250 °C with a heating rate of 0,5 °C/s by two small resistances (each one with a power level of 150 W).

To illustrate the features of such systems, two typical examples of results are presented in the next paragraphs.

3. Millimeter sized particle results: Examples of alumina splat cooling during the first ten milliseconds for an impact with a Weber number close to that obtained with the micrometer sized particle plasma sprayed (We=2500)

Table 1 shows images of 5 mm alumina drops impacting onto the mirror-polished stainless steel substrate at two different temperatures.

On a cold surface (room temperature), the particle spreads out more, and also cools down faster. This cooling down is such that, according to the pyrometer measurements, the central part surface is solidified when the piston arrives at the end of its race (i.e. 14 ms after the impact). Splashing is due to the rupture of the thin film expanded around the central part of the splat and the presence of a matter accumulation at splat edges. After the break of the thin film, there is no splat evolution and the splat cooling is slower.

On a preheated surface, there is no thin film expanded around the splat centre. So it can be supposed that flattening is impeded by the splat solidification at substrate surface and the splashing observed is due to liquid droplets jetting out at its top surface still liquid. At the splat edges it can be observed some matter recoil towards the splat centre. Finally a disk shaped splat is obtained.

Table 1. Flattening of a millimeter sized alumina drop onto a mirror polished stainless steel substrate held at two different temperatures

Time (ms)	0	1	2	3	4	5	10
$T_{substrate}$= 20°C $T_{particle}$= 2100°C $D_{particle}$=5 mm V_{impact}=10m/s		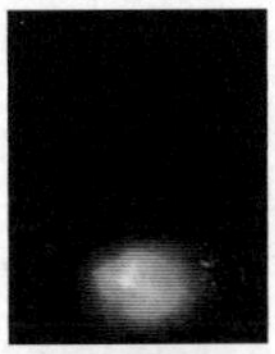	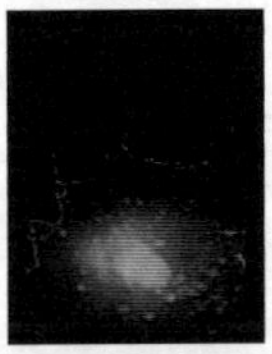	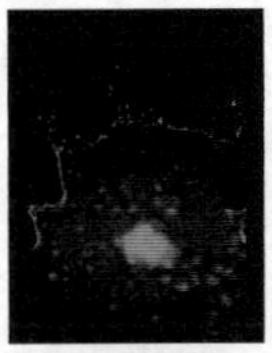	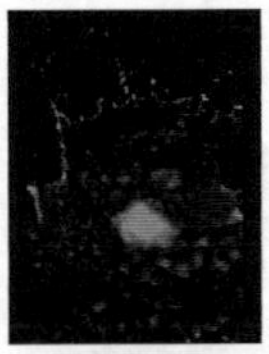	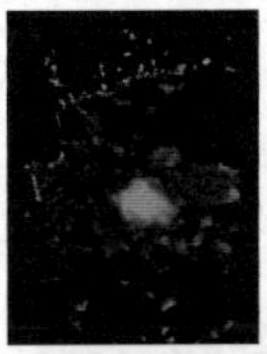	
$T_{substrate}$= 200°C $T_{particle}$= 2100°C $D_{particle}$=5 mm V_{impact}=10m/s	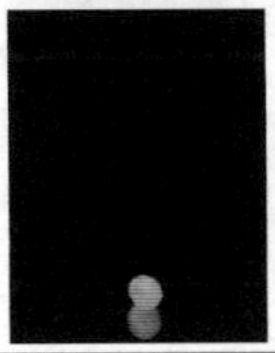	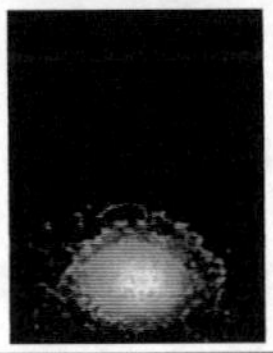	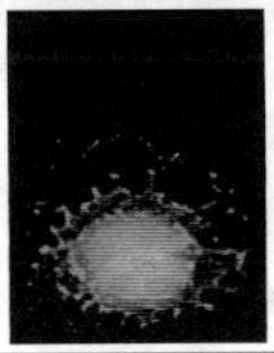				

In order to understand better the differences between both cases, the splat surface temperature is presented in figure 6. It can be seen first that the temperature of the drop is only of 2100°C at the impact, due to properties of the drop generator. Indeed the fall of the drop is only controlled by on the one hand its surface tension and on the other the gravity force. The drop is kept during a long time into the electrical arc and its temperature is near its melting one. The second point that must be underlined is the temperature decreases during the droplet flattening followed by the increase of the temperature up to the melting point. This corresponds to the under-cooling phenomenon.

When both cases are compared, the under-cooling phenomenon seems to last longer on a hot substrate. This effect may be due to the good contact with the substrate, which pumps more energy of the splat inducing slower temperature increase. However an other explanation can be that the splat on a cold surface is much more expanded with more numerous nucleation sites, the thermal energy produced by the nucleation being then more important.

Another point is that the under-cooling phenomenon permits to the splat on a hot substrate to reorganize its edges in order to minimize the surface tension strength and obtain a disk.

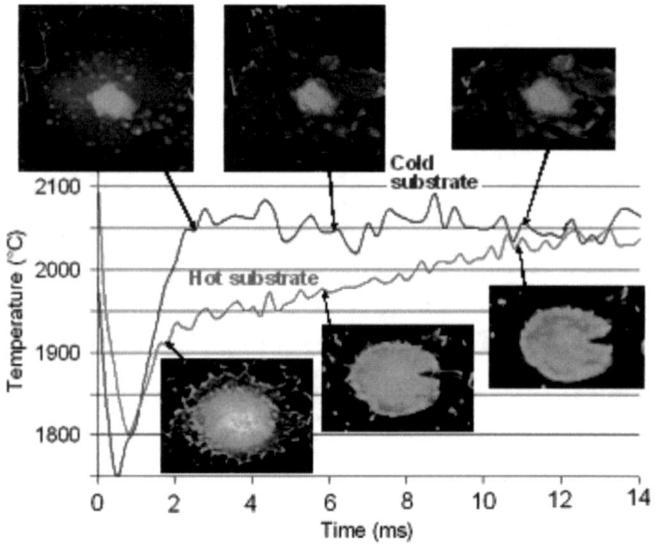

Figure 6.
Superposition of the splat shape and splat surface temperature time evolutions for a millimetre sized alumina drop impacting with a velocity of 10 m/s onto a mirror polished stainless steel substrate held at two different temperatures

4. Micrometer sized drop results: Examples of alumina particle impact at the micrometer size for different preheating conditions.

In figure 7 is presented the impact of an alumina particle on a cold (20°C) stainless steel. In the figure 7 (a) and (b) are presented respectively the views from the camera focusing orthogonally to the substrate and from that focusing tangentially. The exposure time of the first camera is 10 μs, the acquisition starting at the impact. For the second camera, the exposure times are sequenced (open 0,5 μs, close 0,5 μs; t_{total}=10 μs) and the acquisition is starting 1 μs before impact in order to validate the velocity and have an estimation of the particle diameter. Thanks to the camera exposure time, it is possible to observe the all flattening history.

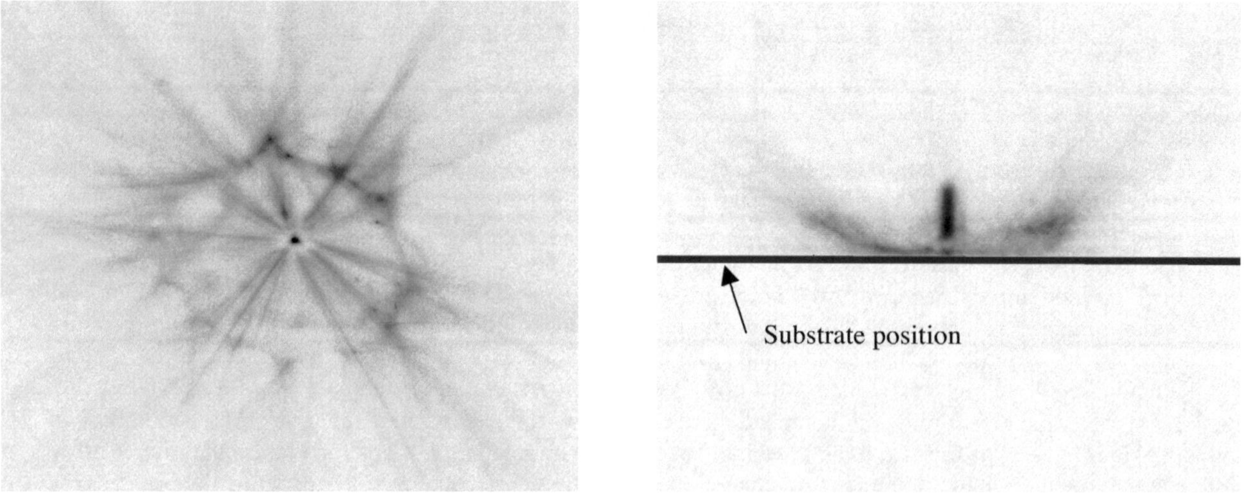

(a) (b)

Figure 7. Flattening of a micrometer sized alumina particle 40 μm in diameter with an impact velocity of 280 m/s onto a mirror polished stainless steel substrate held at room temperature (20°C). Views presented into two directions relatively to the substrate (a) orthogonal (b) tangential.

First it can be noted that the maximum flattening of the drop is about 10 times larger than the initial particle diameter. The second point is that near the impact center, some ejections are present with a starting location circle shaped. This circle corresponds roughly to the diameter of the impacting particle. It is thus clear that impact splashing occurs and induces some matter ejection at impact. Finally between the splat center and its edges a film seems to exist with matter accumulation at splat edges. Peripheral ejections seem to be generated at different locations and not on the whole periphery of the splat.

When ejections at the impact are compared with peripheral ones, in figure 7 (a) it can be seen that the first ejections have radial directions. But the second ejections have random directions. In figure 7 (b), it can be seen that there are many ejection angles, forming an ejection cloud where it is not possible to discriminate the different types of ejections.

Finally, in figure 8 (a) and (b), the resulting splat, at the end of flattening and cooling processes, is presented. It seems that the particle is completely destroyed at impact; the shape of the resulting splat corresponds to a circle around the impact center with a radius of about 200 μm (see figure 8 (a)). From the picture of the camera 1, it can be seen that splashing takes place at splat edges at about a distance of 200 μm from the splat center. So the matter circled around the impact center corresponds to the maximum flattening of the particle and probably to the peripheral ejection area. In the figure 8 (b), a zoom view of the splat center is presenting and it can be seen that it exists some areas without matter, which start from the impact center where there is very few matter. The diameter of the matter remaining corresponds roughly to the initial particle diameter.

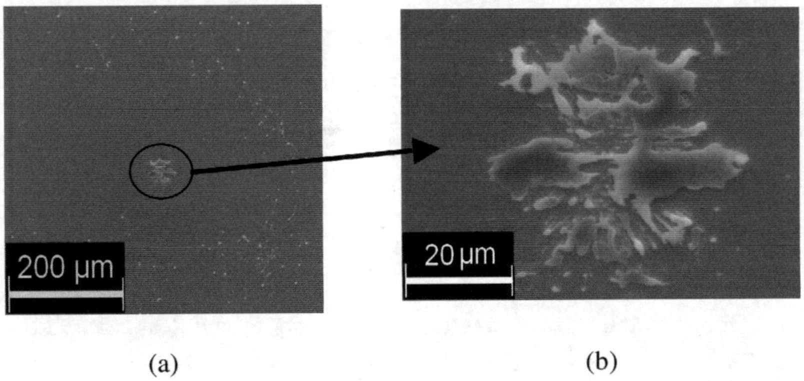

Figure 8. Resulting splat view after flattening and cooling for a particle 40 μm in diameter with an impact velocity of 280 m/s onto a stainless steel substrate held at room temperature (20°C), (a) global view (b) Zoom view of the impact location

On a hot surface (heating 10 min at 250°C) no splashing is observed (see figure 9 (a) and (b)) and the resulting splats are disk shaped with very few thin matter fingers, a thin matter ring being also present at the peripheral (see figure 9 (c)).

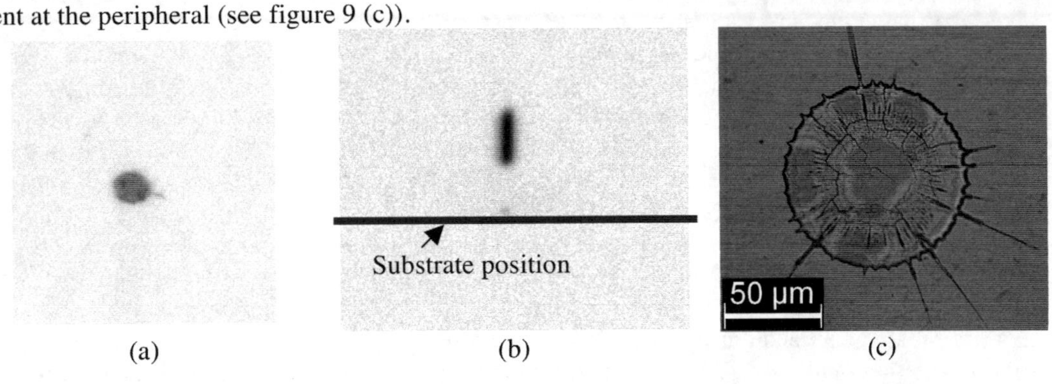

Figure 9. Flattening of a micrometer sized alumina particle 40 μm in diameter with an impact velocity of 290 m/s onto a mirror polished stainless steel substrate held at high temperature (250°C) during 10 min. Views presented into two directions relative to the substrate (a) orthogonal (b) tangential. (c) Resulting splat

In order to understand the difference between these two impacts figure 10 (a) and (b) present the thermal signals from pyrometers and the corresponding calculated temperature in the case respectively of a cold and a hot substrate.

In both cases, during the impact, the particle temperature increases. It is due to the thermal gradient existing inside it. After this temperature increase, the cooling is much more faster onto a hot substrate. The flattening on a cold surface is very fast because the maximum flattening, in spite of the larger diameter, is reached before that obtained on a hot substrate. Indeed in 1.2 μs, the particle on a cold substrate reaches a radius of 200 μm, corresponding to the flattening speed of about of 170 m/s. On the hot substrate, the flattening speed is only to 50 m/s.

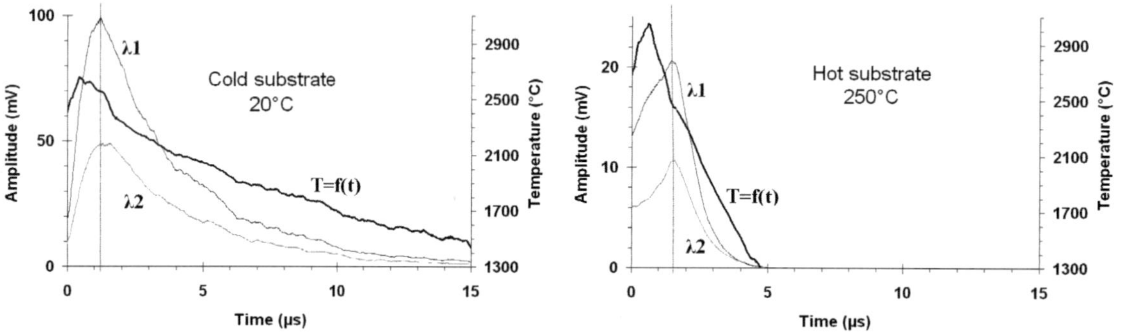

(a) (b)

Figure 10. Thermal signals of the pyrometer generated by the flattening particle surface with the corresponding temperature for two substrate temperatures (a) 20°C (b) 250°C, 10 min. These signals correspond to the particles presented previously.

5. Conclusion

Plasma sprayed coating properties depend strongly on contacts between layered splats. This study is devoted to the understanding of the formation of a single splat on smooth stainless steel substrate according to particle parameters at impact, desorption or not of adsorbates and condensates at substrate surface. As it is not possible to follow the flattening of micrometer sized particle plasma sprayed (occurring in the microsecond time range), measurements have been performed in parallel with millimeter sized ones. The first experimental set-up developed generates millimeter sized (5mm) ceramic or metal drops with a Weber number close to that obtained when plasma spraying particles of the same material in the tens of micrometer size. It permits to follow the particle flattening and cooling during the first 14ms. The second set-up uses a direct current plasma torch with micrometer sized particles that gives global information about the impact of micrometer sized particles onto substrates held at different temperatures. Thanks to a substrate cooling system starting as soon as the splat is formed, it is possible to freeze the oxidation of the substrate and splat surfaces after the particle impact and to study the effect of the substrate surface chemistry on the splat shape. Correlating results obtained with fast (100ns response time) bi-chromatic pyrometer and the imaging techniques it becomes possible to observe the whole flattening history. Results permit to better understand the effect of the substrate surface modifications (oxidation state and corresponding roughness, as well as desorption of adsorbates and condensates) onto the splat shape. This system allows also following the matter ejection at impact and during flattening and solidification. Thus it becomes possible to improve the splat deposition onto the substrate by optimizing its preheating. When results obtained with both systems are compared correlations can be found on the flattening of particles at both scales. It is also possible to compare experimental results with flattening models at both scales and, especially, to better understand the flattening process for micrometer-sized particles and after for approaching nanometer scale. Indeed only a model can be used for observe the flattening of nanometer sized particle, due to the particle size and the very short distance between the torch and substrate.

References

[1] Chandra, S., Fauchais, P., Formation of solid splats during thermal spray deposition, (2009), Journal of Thermal Spray Technology, volume 18 (2), pp. 148-180

[2] Fauchais, P., Fukumoto, M., Vardelle A. and Vardelle M., *Knowledge concerning splat formation*: An invited review, (2003), Journal of Thermal Spray Technology, volume 13 (3), pp. 337-360

[3] Dhiman R, Chandra S, Freezing-induced splashing during impact of molten metal droplets with high weber numbers, (2005), International journal of heat and mass transfer, volume 48, pp 5625-5638

[4] Shinoda K., Murakami H and al, High-speed thermal imaging of yttria-stabilized zirconia droplets impinging on a substrate in plasma spraying, (2007), Appl. Phys. Lett., volume 90, pp 194103/1-194103/3

[5] Mcdonald A, Chandra S., Moreau C, Photographing impact of plasma sprayed particles on rough substrates, (2008), Journal of mater sciences, volume 43, pp 4631-4643

[6] McDonald A, Lamontagne M and al, Photographing impact of plasma-sprayed particles on metal substrates, (2006), Journal of thermal spray technology, volume 14, issue 4, pp 708-716

[7] Goutier, S., Vardelle, M., Labbe, J.C., Fauchais, P., Alumina Splat Investigation: Visualization of Impact and Splat/Substrate Interface for Millimeter-Sized Drops, (2010), Journal of thermal spray technology, Volume 19 (1), pp 49-55

Journal of Physics: Conference Series **275** (2011) 012004 doi:10.1088/1742-6596/275/1/012004

Experimental assessment of the surface temperature of copper electrodes submitted to an electric arc in air at atmospheric pressure.

R. Landfried, T. Leblanc, R.Andlauer, Ph. Teste
Laboratoire de Génie Electrique de Paris : SUPELEC – CNRS – Universités Paris VI et Paris XI - Plateau de Moulon – 91192 Gif sur Yvette Cedex - FRANCE

E-mail: teste@lgep.supelec.fr

Abstract. This paper concerns the assessment of the surface temperature of copper electrodes submitted to an electric arc in a non stationary regime in air. An infrared camera is used to measure the decrease of the temperature surface just after a controlled and very fast arc extinction. In the first part, the experimental method is described. In the second part, results are presented for 60-70 A with an electric arc duration in the range 3-4 ms. The temperature decrease after the arc extinction allows to reach an assessment of the surface temperature just at the arc switching off. In the present experimental conditions the mean temperatures reached for copper cathodes and anodes are in the range 750 – 850°C.

1. Introduction

Understanding of arc-electrode interaction in air at atmospheric pressure is still incomplete because of the lack of information concerning the evolution of surface characteristics during the arc. Among these important informations, the surface temperature remains in the case of non stationary electric arcs in air an experimentally relatively unknown parameter. Many researches have been conducted in the case of stationary arcs burning in argon with refractory electrodes (tungsten, thoriated tungsten, hafnium ...) [1]-[3]. In these cases, the methods used to measure surface temperature depend strongly on the stationary character of the arc. In [1]-[5] optical methods were used such as single color and a double color pyrometry. The stability of the arc root on the electrode surface and the arc column symmetry were often an essential parameter to allow the temperature measurement.

This paper deals with the surface temperature distribution study on copper (low melting temperature) anodes and cathodes in the case of a switching arc (non stationary regime) burning in air at atmospheric pressure with an electric arc current around some tenths of amperes. In this case the use of a pyrometric method is difficult. Indeed, the arc may be very unstable on the electrode surface, and then no symmetry considerations can be used and a very short integration time for the measurement is then required.

The paper is organized as follows:

- In the first part, the experimental device used to create opening electric arcs is described.
- In the second part, the experimental method used to measure the temperature is presented and discussed.
- In the third part, we present results concerning copper anode and cathode surface temperature measurements at different instant after the arc extinction. An assessment of the surface temperature at the time of the arc extinction is also proposed.

2. Description of the experimental device

The experimental device used to measure the temperature of the electrode surface may be divided into two sets:

- the electrical device which consists of the power supply, the load circuit and the switching off electronic system. It is presented in figure 1.

- the electromechanical system which opens the contact and the different cameras used to observe the arc behavior and to measure the surface temperature. It is schematically presented in figure 2.

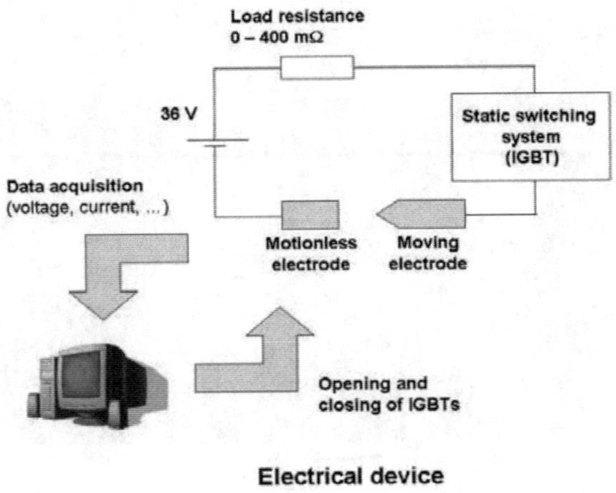

Figure 1. Schematic description of the electrical part of the experimental device

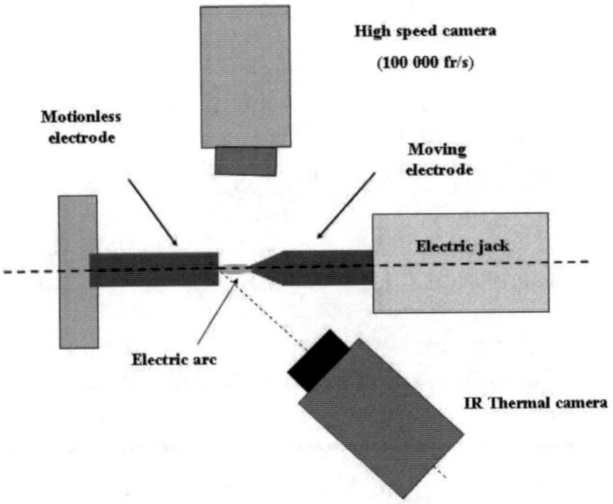

Figure 2. Schematic description of the experimental device.

2.1 The electrical device

The power supply is made up with a set of three 12V electric batteries. The internal resistance of this set is about 18 mΩ. An electronic device, made of IGBTs (Insulated Gate Bipolar Transistors) allows the control of the intensity pulse in the circuit and is also used to switch off the arc current. This static device may switch off a current intensity as high as 1500 A in less than a microsecond and may dissipate up to 275 J. The load in the circuit is made of an adjustable resistance (0 to 400 mΩ) and is supposedly no inductive.

2.2 The mechanical device

Two coaxial electrodes constitute the contact. A motionless one is a cylinder and has a flat surface. In order to increase the observable area on the flat electrode when the contact is open, a conical shape was chosen for the moving electrode.

Journal of Physics: Conference Series **275** (2011) 012004 doi:10.1088/1742-6596/275/1/012004

A photograph presented in figure 3 and taken before the arc illustrates the position and the shapes of the two electrodes. The electric jack is used to open the contact with nearly constant velocity.

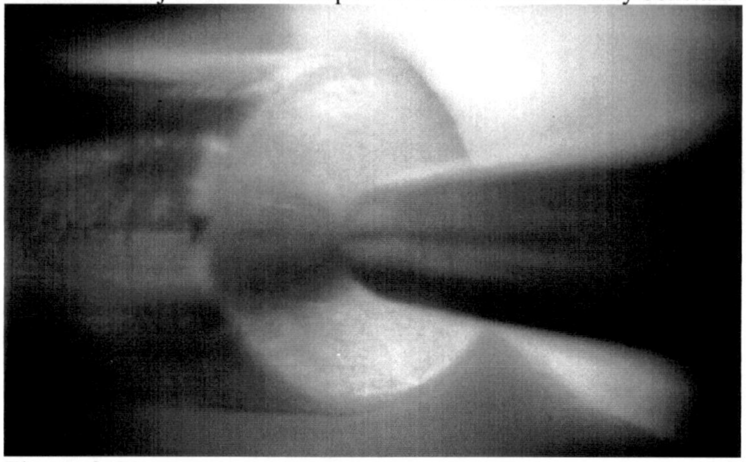

Figure 3. Photography of the electrodes before the arc: the contact is closed.

3. Measurements and description of the method

During the electric arc the following parameters were recorded:

- The current intensity.
- The arc voltage.
- The electrode gap.

Moreover, the electric arc was filmed with the help of a high speed camera (100 000 frames/s). To measure the temperature distribution on the flat motionless electrode surface, an infrared camera was used. The spectral response of the InSb camera detector is in the range 3.7 μm – 4.8 μm. The matrix detector is 320 pixels x 240 pixels. The optic used allows a spatial resolution of 40 μm to 60 μm by pixel. The first operation to achieve was to calibrate this camera for copper electrodes. It was realized with the help of a specific device described in [6] which allowed to calibrate precisely the camera until electrode fusion temperature.

Because of the intense light of the electric arc it was not possible to photograph the electrode surface and obtain the surface temperature distribution during the arc duration. The electronic system allowed to force the switching off of the electric arc in less than 1 μs (this duration is denoted τ_{elec}).

The control electronic developed with the infrared camera allowed to activate this one with a time delay of 15 μs (this duration is denoted $\tau_{activate}$) and has a precision less than 0.1 μs. The exposure time of the infrared camera is denoted $\tau_{exposure}$. According to the temperature range, $\tau_{exposure}$ takes values in the range [10 μs, 120 μs].

Then, the duration denoted τ_{image} will be considered as the instant for which the photography is taken and is such as:

$$\tau_{image} = \tau_{elec} + \tau_{activate} + \frac{1}{2}\tau_{exposure}$$

These different times are schematically presented in figure 4.

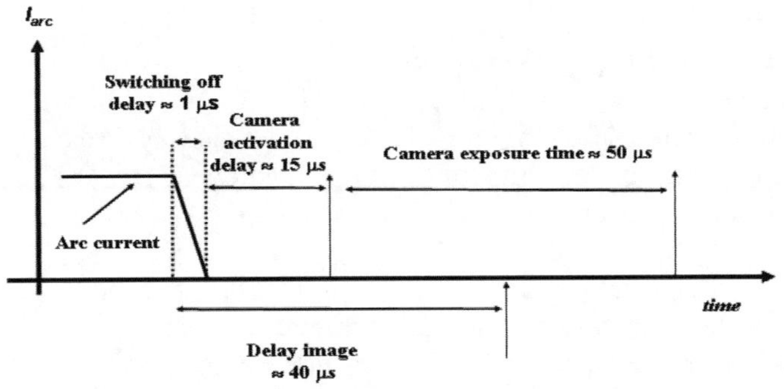

Figure 4. Description of the different delays - $\tau_{exposure}$ = 50 µs.

The electrode material was copper OFHC (Oxygen Free High Conductivity). The covering gas was air at atmospheric pressure. The diameter of the electrodes was 8 mm. The conical electrode has a half angle at the top equal to 30°. Moreover, an electric jack was chosen in order to allow a precise synchronization (<0.1 µs) between the motion and the IR camera. Consequently, this choice limited the opening velocity.

The process of opening operation is as follows:

When the IGBTs are closed, DC current in the electrical circuit is in the range 100 A – 150 A. When the contact opens the arc current drops rapidly and becomes stable around its mean value (≈ 60/70 A) and decreases slowly during the arc until the switching off. At this instant, its value is in the range 50 A – 60 A. The contact voltage increases at the contact opening to the arc voltage value and remains quite stable during the arc. The arc duration is controlled by the IGBTs triggering which depends on the software control for which exists an uncertainty of 2 ms. For each opening, the data are automatically recorded. Figure 5 presents typical evolution for contact voltage, current intensity and electrodes gap during the contact opening.

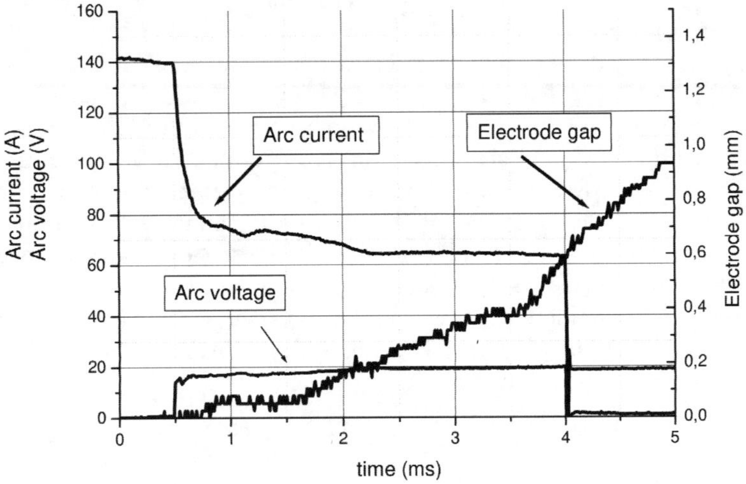

Figure 5. Typical time evolution of arc voltage, arc current and electrode gap.

4. Experimental results – Assessment of the surface temperature

The temperature profiles were investigated for:

- an average arc current $I_{arc} \approx 70$ A
- arc durations dt in the range 3 ms - 4 ms
- various values of τ_{image} (40 µs, 200 µs, 500 µs and 1 ms) in order to observe the temperature decrease after the arc extinction and then to propose an assessment of the surface temperature at the instant of the arc extinction.

For each τ_{image}value about ten measurements were made.

Preliminary remarks:

- The choices of I_{arc} value and of the arc durations have been imposed by the present limitation in the camera calibration temperature range. For higher arc current intensity or for longer electric arcs, temperatures reached on the surface may be greater than the melting temperature.
- It should be emphasised that temperature distribution observed using thermal mapping can not show the heating resulting from the existence of microscopic structures such as fragments or microspots observed in [7] which have life durations equal respectively to 20 ns and 10 µs and a spatial size of some micrometers. Indeed, the exposure time of the infrared camera is quite greater than the life duration of these structures and the size of an observed pixel is quite greater than a microspot size.

In these conditions the temperature distributions obtained show only an average temperature distribution that may result from a multi spot intense heating that may be represented by the heating of a single macroscopic arc root [8].

In the first part temperature measurement for copper anodes are presented for different values of τ_{image} in the range 40 µs – 1000 µs.

The second part concerns results for copper cathodes and the third part concerns the assessment of the surface temperature at the extinction instant.

4.1 Results concerning copper anodes

4.1.1 Results for τ_{image} = 40 µs.

In this case $\tau_{exposure}$ = 50 µs (and $\tau_{activate}$ = 15 µs). In this case the mean arc duration was equal to 3.6 ms, the average arc current was equal to 68 A. The mean maximum temperature reached on the anode surface (T_{max}) was equal to 750 °C.

The values of arc durations and the average arc currents are quite close for all seven cases, however an important dispersion in the T_{max} values was observed. Different explanations may be proposed for that:

- images obtained with the high speed camera have shown that the arc mobility on the electrode surface (and the arc stagnation time) could be different from one electric arc to another one. Moreover, in two cases no stagnation time could be determined: the electric arc seems to move by jumps or continuously during all the arc duration.
- The arc current intensity at the switching off instant may vary from one arc to another between 48 and 62 A. This dispersion may also be due to the different arc behaviors.

Two surface temperature distributions are presented in figures 6(a) and 6(b). In figure 6(a) the electric arc was motionless some ms before the switching off. A single heated area is discernible. The temperature distribution is practically symmetric.

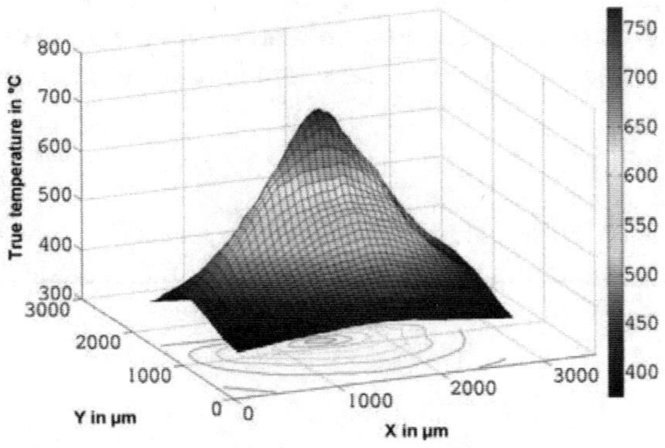

Figure 6(a). Example of surface temperature distribution (copper anode) in the case of a motionless electric arc.

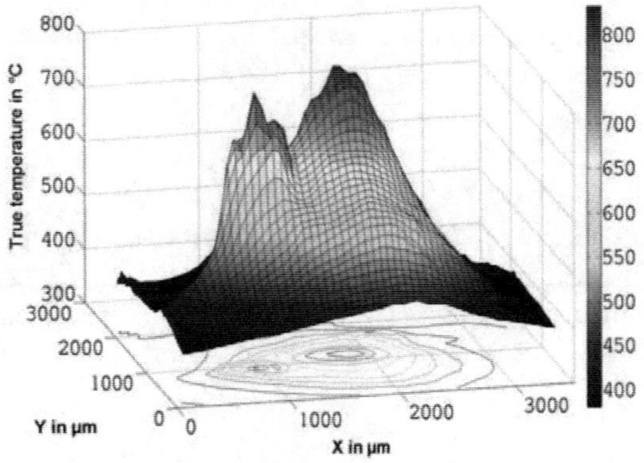

Figure 6(b). Example of surface temperature distribution (copper anode) in the case of a jumping arc root during the arc.

On the contrary in figure 6(b) the electric arc has moved just before switching off. Two heated zones are discernable. In the case of nearly motionless arcs (at the end of the arc), a typical profile of temperature taken along an arc root diameter is plotted in figure 7.
In all the observed cases, the temperature peak width for T = 500°C is less than 1.5 µm and for T = 600°C it is less than 900 µm. The maximum surface temperature gradient value obtained is around 440 °C/mm.

4.1.2 Results for τ_{image} = 200 µs
In this case $\tau_{exposure}$ = 50 µs (and $\tau_{activate}$ = 15 µs). In this case the mean arc duration was equal to 3.5 ms, the average arc current was equal to 72 A. The mean maximum temperature reached on the anode surface (T_{max}) was equal to 620 °C. In figure 8 a typical profile is plotted. The maximum surface temperature gradient value obtained is around 310 °C/mm.

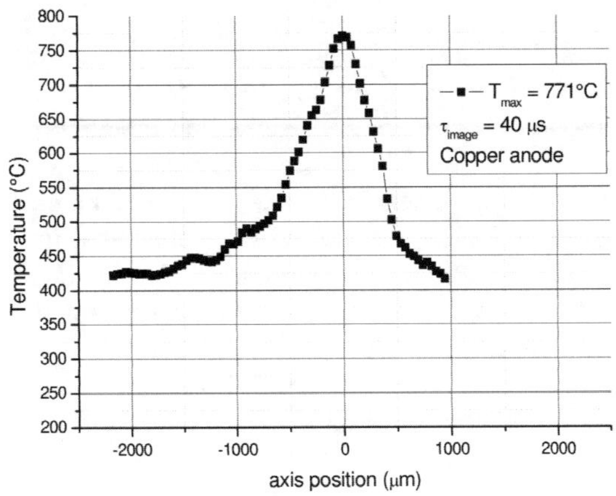

Figure 7. Typical surface temperature profile along the diameter of the arc root for τ_{image} = 40 µs. Case of copper anode.

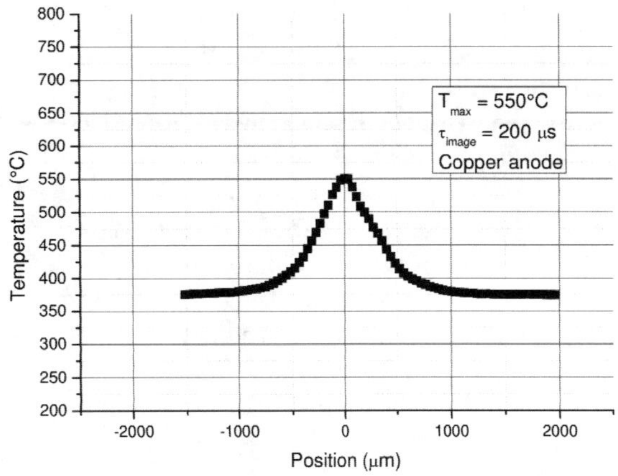

Figure 8. Typical surface temperature profile along the diameter of the arc root for τ_{image} = 200 µs. Case of copper anode.

4.1.3 Results for τ_{image} = 500 µs

In this case $\tau_{exposure}$ = 10 μs (and $\tau_{activate}$ = 15 μs). In this case the mean arc duration was equal to 3.5 ms, the average arc current was equal to 69 A. The mean maximum temperature reached on the anode surface (T_{max}) was equal to 398 °C. A typical value of the maximum surface temperature gradient value obtained is around 235 °C/mm.

4.1.4 Results for τ_{image} = 1000 μs.
In this case $\tau_{exposure}$ = 10 μs (and $\tau_{activate}$ = 15 μs). In this case the mean arc duration was equal to 3.7 ms, the average arc current was equal to 72 A. The mean maximum temperature reached on the anode surface (T_{max}) was equal to 322 °C. A typical value of the maximum surface temperature gradient value obtained is around 175 °C/mm.

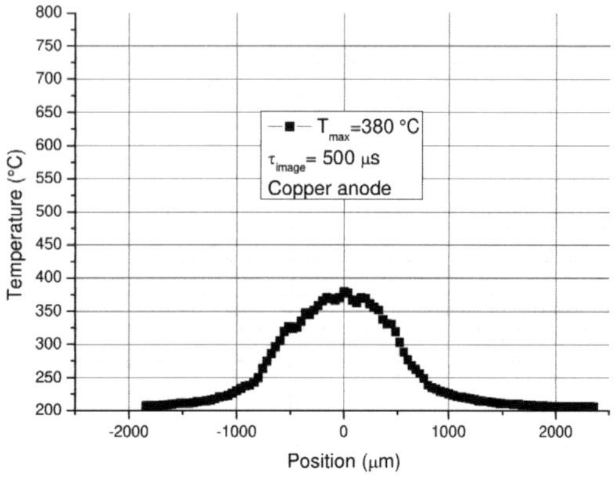

Figure 9. Typical surface temperature profile along the diameter of the arc root for τ_{image} = 500 μs. Case of copper anode.

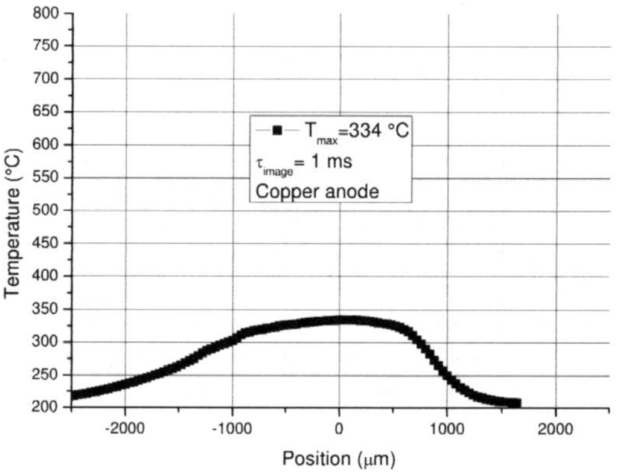

Figure 10. Typical surface temperature profile along the diameter of the arc root for τ_{image} = 1000 μs. Case of copper anode.

4.2. Results concerning copper cathodes
4.2.1 Results for τ_{image} = 40 μs

In this case $\tau_{exposure}$ = 50 μs (and $\tau_{activate}$ = 15 μs). In this case the mean arc duration was equal to 3.5 ms, the average arc current was equal to 63 A. The mean maximum temperature reached on the anode surface (T_{max}) was equal to 710 °C.

- Two examples of surface temperature distribution are proposed in figures 11(a) and 11(b) and two different typical temperature profiles along the arc root diameter are presented in figure 12. Two kinds of profile appear in the case of cathode: one is quite similar to the temperature profiles observed for anodes. In these cases, the maximum surface temperature gradients that have been observed may reach 580°C/mm.
- Another type of profile does not show a significant temperature peak, it is wider and has a "flat" top. That may be due to a continuous motion of the arc root on the cathode just at the end of the electric arc.

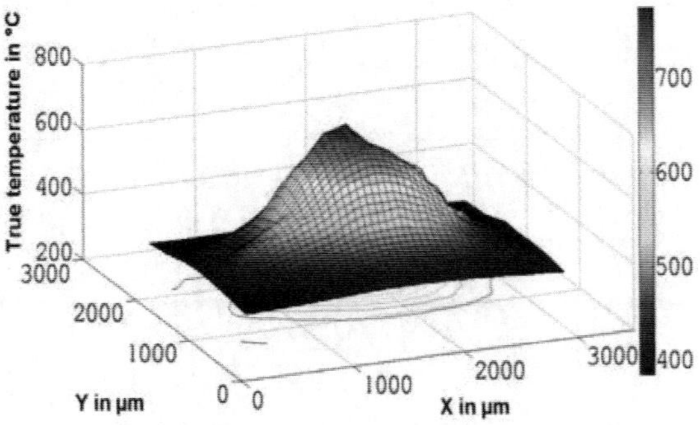

Figure 11(a). Example of surface temperature distribution (copper cathode) in the case of a motionless electric arc.

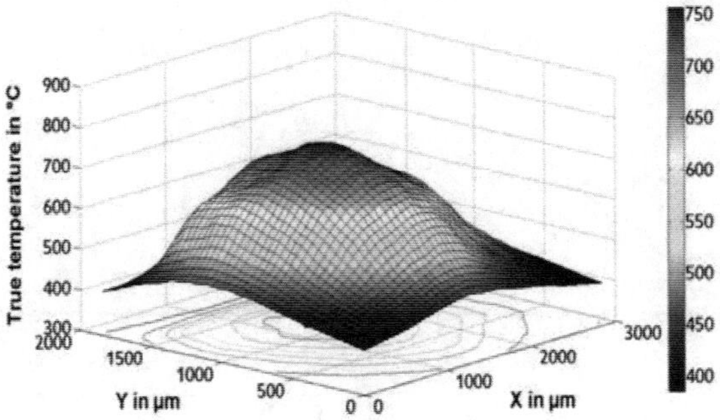

Figure 11(b). Example of surface temperature distribution (copper cathode) with flat top.

4.2.2 Results for τ_{image} = 200 µs

In this case $\tau_{exposure}$ = 50 µs (and $\tau_{activate}$ = 15 µs). In this case the mean arc duration was equal to 3.5 ms, the average arc current was equal to 71 A. The mean maximum temperature reached on the anode surface (T_{max}) was equal to 600 °C. In figure 13 a typical profile is plotted. The maximum surface temperature gradient value obtained is around 300 °C/mm.

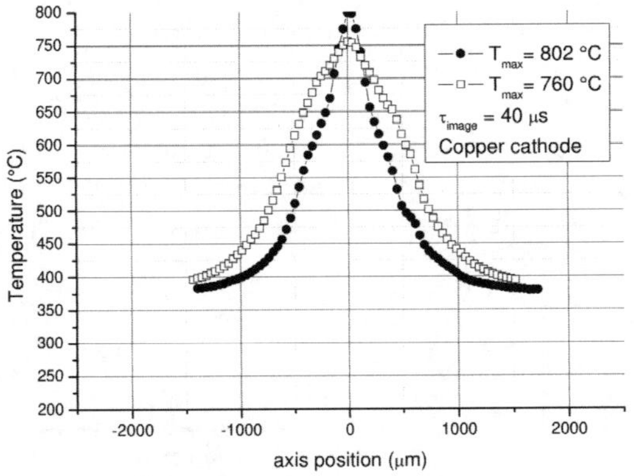

Figure 12. Typical surface temperature profiles along the diameter of the arc root for τ_{image} = 40 µs. Case of copper cathodes.

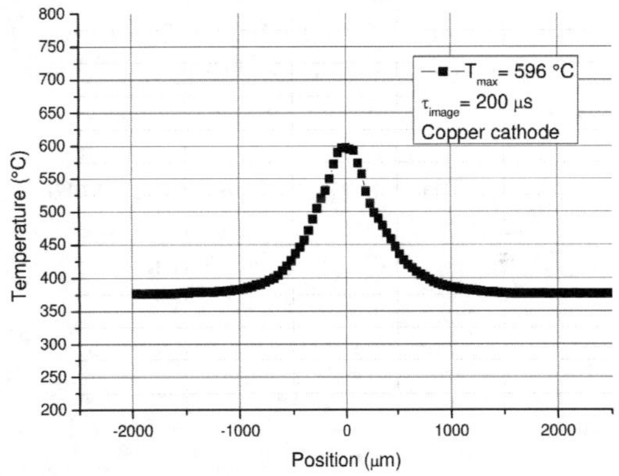

Figure 13. Typical surface temperature profiles along the diameter of the arc root for τ_{image} = 200 µs. Case of copper cathodes.

4.2.3 Results for τ_{image} = 500 µs.

In this case $\tau_{exposure}$ = 10 µs (and $\tau_{activate}$ = 15 µs). In this case the mean arc duration was equal to 3.5 ms, the average arc current was equal to 72 A. The mean maximum temperature reached on the anode surface (T_{max}) was equal to 375 °C. In figure 14 a typical profile is plotted. The maximum surface temperature gradient value obtained is around 180 °C/mm.

4.2.4 Results for τ_{image} = 1000 µs

In this case $\tau_{exposure}$ = 10 µs (and $\tau_{activate}$ = 15 µs). In this case the mean arc duration was equal to 3.6 ms, the average arc current was equal to 73 A. The mean maximum temperature reached on the anode surface (T_{max}) was equal to 330 °C. In figure 15 a typical profile is plotted. The maximum surface temperature gradient value obtained is around 130 °C/mm.

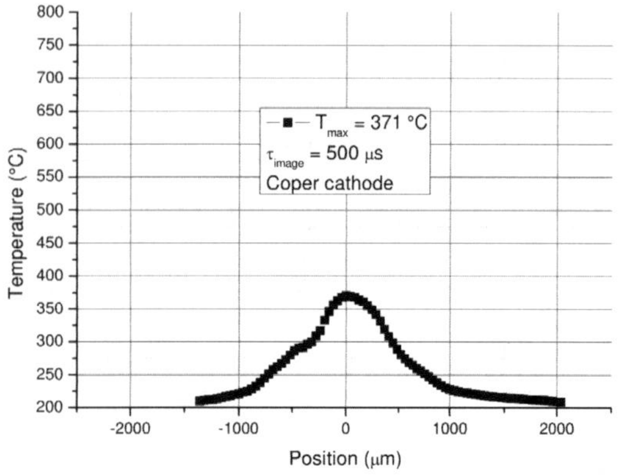

Figure 14. Typical surface temperature profiles along the diameter of the arc root for τ_{image} = 500 μs. Case of copper cathodes.

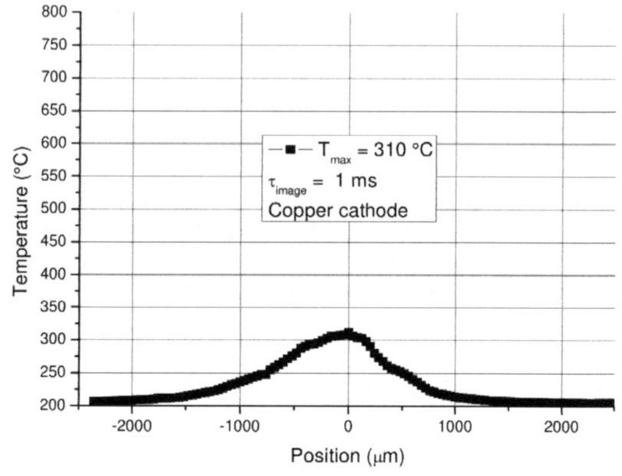

Figure 15. Typical surface temperature profiles along the diameter of the arc root for τ_{image} = 1000 μs. Case of copper cathodes.

4.3 Assessment of the surface temperature at the time of the arc extinction

Measurements made give the surface temperature 40 μs after the time of the arc extinction. With the help of the temperature evolution with τ_{image} it is possible to propose an extrapolation of the surface temperature and then obtain the surface temperature at the extinction time of the electric arc.

In figures 16 and 17 the evolution of the maximum temperature versus τ_{image} is plotted for anode and cathodes respectively. The dispersion is also presented. In [9] it has been shown that for a 1D problem and for very similar thermal conditions the temperature decrease with time was exponential. Two exponential fits have then also been plotted in these figures. They allow to reach an assessment of the surface temperature at the time of the arc extinction. A mean value of 815°C was obtained for copper anodes and 778°C for copper cathodes.

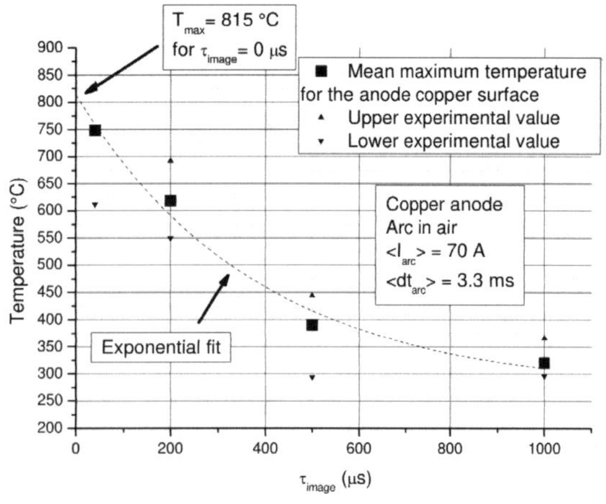

Figure 16. Evolution of the maximum surface temperature versus τ_{image} for a copper anode. Exponential fit of the curve and assessment of the surface temperature at the time of the arc extinction.

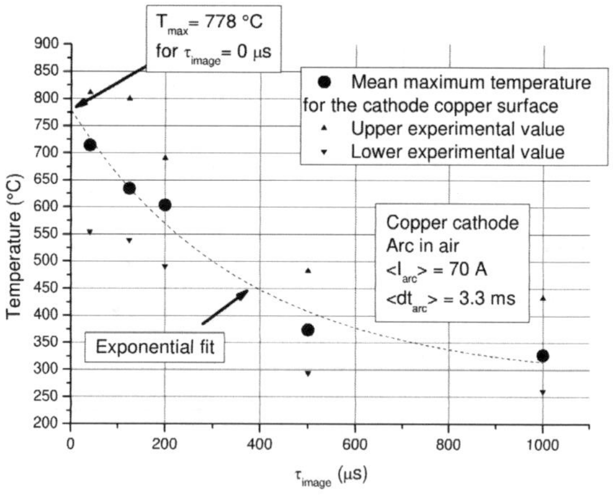

Figure 17. Evolution of the maximum surface temperature versus τ_{image} for a copper cathode. Exponential fit of the curve and assessment of the surface temperature at the time of the arc extinction.

It should be emphasized that the difference between the surface temperature at τ_{image} = 40 μs and τ_{image} = 0 μs may reach more than 50 °C.

5. Conclusions.

The aim of this paper was to obtain information concerning the surface temperature of copper anodes and cathodes submitted to a non stationary (average duration of 3.5 ms) electric arc burning in air at atmospheric pressure. For that the temperature distribution on the electrode surface was measured with a short delay of 40 μs after the arc extinction. The experimental method brought into operation presents several advantages:

- The IR camera may be used with short exposure times (between 10 and 120 μs)
- The duration of the switching off of the electric arc is forced and brutal (more than 200 A/μs).
- The IR camera may be precisely synchronized.

For such experimental conditions (non stationary arc in air, copper electrodes, ...) it was then possible to obtain for the first time a temperature map of the electrode surface 40 μs after the arc extinction.
For copper anodes, only one sort of temperature distribution was observed and different types of temperature profiles appeared for copper cathodes. The observation made with the help of a high speed camera allowed to think that it was due to various arc behaviors (stagnation, jumps or continuous motions).
The measurement of the temperature distribution for different delays allowed also to propose by extrapolation an assessment of the temperature at the arc extinction instant. Maximum average temperatures of 815°C for the copper anode and 780°C for the cathode seem to be good assessments.

6. References

[1] J. Haidar and J D Farmer 1993 *J. Phys. D: Appl. Phys.* **26** 1224-1229
[2] J. Haidar and J D Farmer 1995 *J. Phys. D: Appl. Phys.* **28**, 2089-2094
[3] X Zhou, B. Ding and J. Heberlein, 1995 *Proc. Holm Conference* 219-231
[4] L. Dabringhausen, D. Nandelstädt, J. Luhmann and J. Mentel 2002 *J. Phys. D: Appl. Phys.* **35** 1621-1630
[5] J. Peters, F. Yin, C. F M Borges, J. Heberlein and C. Hackett 2005 *J. Phys. D: Appl. Phys.* **38** 1781-1794
[6] T. Klonowski 2007 PhD Thesis University Paris Sud 11
[7] B. Jüttner 2001 *J. Phys. D: Appl. Phys.* **34** R103-R123
[8] Ph. Teste, J. Rossignol, 2008 *J. of High Temperature Material Processes* **12** Issue 1 43-58
[9] Carslaw and Jaeger 1959 Conduction of heat in solids, Oxford University Press

Journal of Physics: Conference Series **275** (2011) 012005 doi:10.1088/1742-6596/275/1/012005

Plasma-arc reactor for production possibility of powdered nano-size materials

V Hadzhiyski[1], M Mihovsky[2], R Gavrilova[3]

University of Chemical Technology and Metallurgy-Sofia, Plasma Metallurgy Research Laboratory "PLASMALAB"
plasmalab@uctm.edu[1], mihovsky@uctm.edu[2], r.gavrilova@abv.bg[3]

Abstract. Nano-size materials of various chemical compositions find increasing application in life nowadays due to some of their unique properties. Plasma technologies are widely used in the production of a range of powdered nano-size materials (metals, alloys, oxides, nitrides, carbides, borides, carbonitrides, etc.), that have relatively high melting temperatures. Until recently, the so-called RF-plasma generated in induction plasma torches was most frequently applied [1-3].

The subject of this paper is the developments of a new type of plasma-arc reactor, operated with transferred arc system for production of disperse nano-size materials. The new characteristics of the PLASMALAB reactor are the method of feeding the charge, plasma arc control and anode design.

The disperse charge is fed by a charge feeding system operating on gravity principle through a hollow cathode of an arc plasma torch situated along the axis of a water-cooled wall vertical tubular reactor. The powdered material is brought into the zone of a plasma space generated by the DC rotating transferred plasma arc.

The arc is subjected to **A**uto-**E**lectro-**M**agnetic **R**otation (AEMR) by an inductor serially connected to the anode circuit.

The anode is in the form of a water-cooled copper ring. It is mounted concentrically within the cylindrical reactor, with its lower part electrically insulated from it.

The electric parameters of the arc in the reactor and the quantity of processed charge are maintained at a level permitting generation of a volumetric plasma discharge. This mode enables one to attain high mean mass temperature while the processed disperse material flows along the reactor axis through the plasma zone where the main physico-chemical processes take place. The product obtained leaves the reactor through the annular anode, from where it enters a cooling chamber for fixing the produced nano-structure.

Experiments for AlN synthesis from aluminium power and nitrogen were carried out using the plasma reactor described here above.

1. Introduction

During last several years PLASMALAB (Plasma Metallurgy Research Laboratory) developed a new construction of plasma arc reactor for reduction processing of raw and waist materials [5-8], based on Bethlehem's "Falling film" (FFP)-reactor [4, 5].

Figure 1 FFP-plasma reactor „PLASMALAB" with charge feeding system [7-9]

The development of the FFP–reactor idea, according to PLASMALAB's concept, involves mainly **the method of feeding the processed raw materials and the plasma arc organization** [6-9].

In the PLASMALAB FFP-reactor (figure1) charge particles are sucked by the rotating arc at the exit of the hollow graphitic cathode, and are transferred to the wall of the concentrically mounted tubular reactor-anode [6-10]. It is assumed that the particles will be heated and melted in the volume of the arc column, while depositing on the reactor-anode wall. The melted metal-slag suspension flows down the reactor anode wall and reduction processes run. The obtained metal-slag melt drops into the rector tundish, where final reduction and gravitational separation of the two liquids take place.

Preliminary experiments with the designed and manufactured laboratory FFP-plasma reactor PLASMALAB (figure 1) [6-10] showed that its productivity and efficiency of processing the powder charge depends mainly on the way of organizing the plasma arc rotation and on arc geometrical dimensions.

Figure 2 shows a scheme of the original "PLASMALAB" system for **A**uto-**E**lectro-

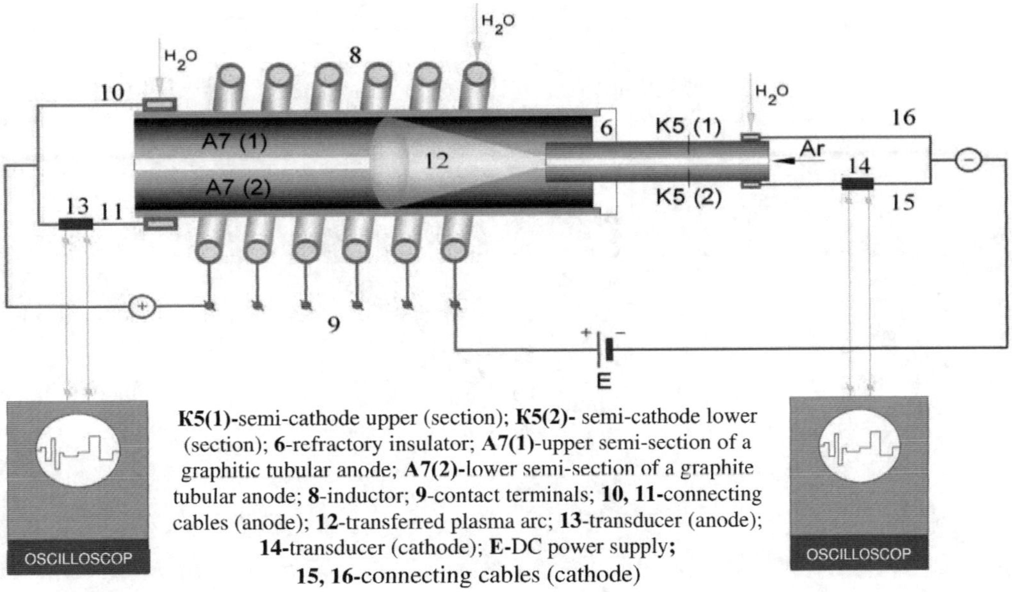

K5(1)-semi-cathode upper (section); **K5(2)**- semi-cathode lower (section); **6**-refractory insulator; **A7(1)**-upper semi-section of a graphitic tubular anode; **A7(2)**-lower semi-section of a graphite tubular anode; **8**-inductor; **9**-contact terminals; **10, 11**-connecting cables (anode); **12**-transferred plasma arc; **13**-transducer (anode); **14**-transducer (cathode); **E**-DC power supply; **15, 16**-connecting cables (cathode)

Figure 2 Scheme of a stand for the experimental study of the rotation speed of the anode and cathode spots of a plasma arc in FFP-plasma reactor "PLASMALAB" [8]

Magnetic-**R**otation (AEMR) of plasma arc [6]. The DC transferred plasma arc is generated by the current that runs in the inductor. This figure illustrates also the idea of finding the rotating speed of the anode and cathode spot [8].

2. "Rotating arc column" and "arc volume"

An interesting phenomenon is observed during the experimental determination of the arc rotation speed [9]. Consider a reactor "PLASMALAB" operating in "idle regime" (i.e. without supply of material for processing) and under fixed arc current value. Then, an arc with a constant length performing an Auto-Electro-Magnetic Rotation in the reactor practically keeps constant revolutions during a specific period of time (in this specific case - about 1÷5 min after the start of arc burning). After the expire of that time, the transducer abruptly stops registering rotation of the cathode spot, and in a while it stops registering the anode spot rotation, too, i.e. it seems that "***the arc stops rotating***".

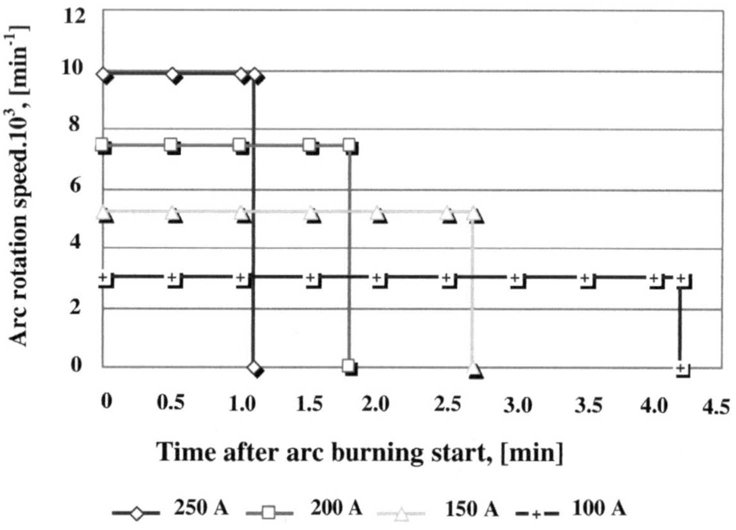

Figure 3 Change of arc rotation speed for the time of constant arc current values 100, 150, 200 and 250 A

As found, that effect is accompanied by change of the noise produced by the burning arc. When arc spots rotate, the sound released is louder than that released by a freely burning arc (without electro-magnetic control). Moreover, sound pulsations are registered for low current values. At the moment when "***the arc stops rotating***", intensity of arc sound and pulsation notably decreases.

To throw light on that phenomenon, we perform control of arc spot rotation in time considering different values of arc current, while the rest of the arc parameters are kept constant. As seen in figure

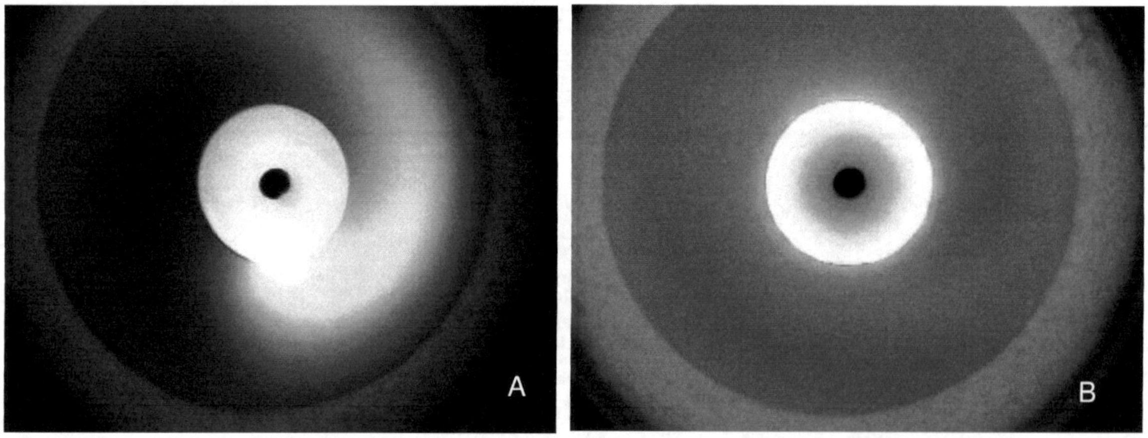

Figure 4 Plasma Arc in FFP-Reactor at different mode of burning
A - Spiral motion of the arc column; B - Plasma volume

7, the effect "***termination of arc rotation***" occurs faster when higher current is applied. Consider current of 100 A. Then, arc rotation is observed for 4.2 min at the average. Consider now current of 250 A. Then, arc rotation is not observed after 1.1 min at the average. Visually, arc rotation ceases long before counters (13 and 14 in figure 2) stop registering impulses, which are generated by rotation of the anode and cathode spot, respectively. Another observed phenomenon is that cathode spot rotation ceases earlier than anode spot rotation. The following hypothesis is formulated as a result of measurements performed and phenomena observed.

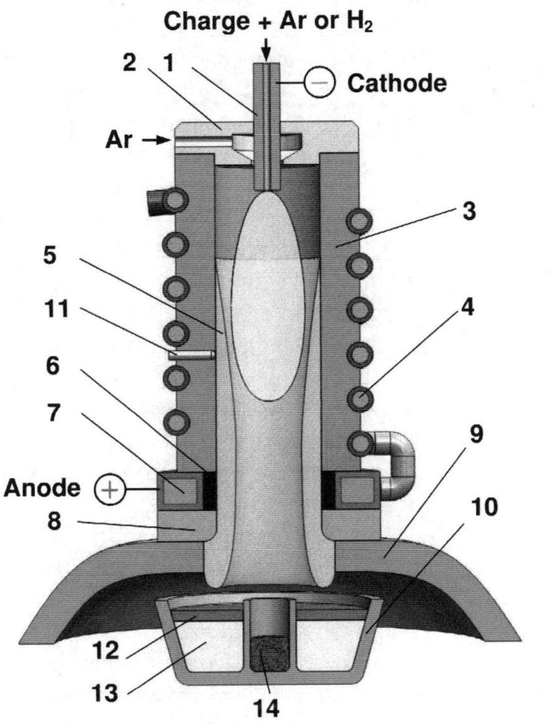

Figure 5 Scheme of the improved plasma reactor "PLASMALAB"
1 – hollow graphite cathode; 2 - nozzle; 3 –concrete reactor; 4 – water-cooled inductor serially connected in the anode chain; 5 – skull layer; 6 – graphite bush-anode; 7 – contact clamp; 8 – refractory insulator; 9 – tundishe roof; 10 – special design graphite crucible; 11 – thermocouple; 12 – slag; 13 – metal; 14 – unprocessed charge.

The arc, having significant power, burns in an arbitrary small space bounded by the wall of the tubular reactor-anode. The temperature of the reactor working surface, even when applying water cooling, rises comparatively quickly up to values higher than 1500-1600 ^{0}C (if no material is fed for processing, this takes several minutes). At such temperature and under high speed of rotation of the arc column, gas volume in the reactor is ionized to a degree which provokes the generation of a number of additional gas electric discharges outside the basic arc column. At a specific moment, thanks to the high temperature, i.e. to the increased degree of ionization, the **arc column** looses its characteristic contours and transforms into a **plasma volume**, which fills the reactor working space.

Considering further technological development of the design of the plasma reactor "PLASMALAB", it is extremely important to find out how the transformation of the rotating arc column into plasma volume will affect the effective transport of powder charge from the aperture of the hollow graphitic cathode to the wall of the tube reactor-anode. One should also find the efficiency of the heat-exchange running between the powder charge and the plasma volume.

Experiments of processing the powder charge (a mixture of converter slam, fine coke and fluxes) are performed for that purpose. The charge particle size is lower than 0.1 mm, and charge is uniformly fed into the improved plasma reactor "PLASMALAB" [9] by means of the powder-feeding system "PLASMALAB" [7].

Some modifications of the initial idea of the design and operation of the system "plasma torch with hollow graphite cathode - water-cooled copper tubular reactor anode" were made on the basis of numerous preliminary experiments [6, 8, 11-13], results found and analysis of a number of undesirable effects. These structural modifications of the improved "PLASMALAB" reactor are shown in figure 5.

A graphite crucible of special design 10 is placed along the axis of reactor 3 (figure 5), at its lower end, under the water cooled graphite anode 6 and in the tundishe. Products of the pyro-metallurgical process are collected in it. These are metal-slag melt and eventually unprocessed charge, unmelted and not transported by the arc to the reactor wall.

Copper ring-wise water-cooled inductor (figure 5) plays the role of a classical copper water-cooled reactor. Reactor space is walled by refractory concrete, and the inductor serves as reinforcement. Thus, thermal losses in the reactor are drastically decreased. Note that the reactor wall is already electrically non-conducting, and the anode power supply is connected to the graphite bush-anode through the inductor. Hence, maximal length of the burning arc is naturally attained.

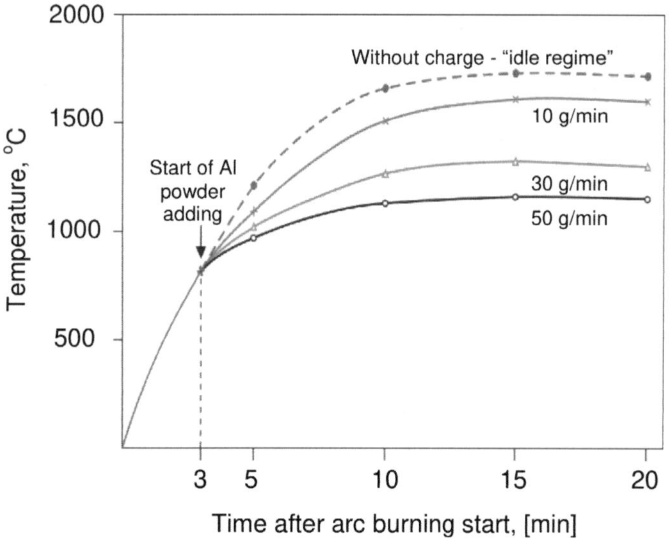

Figure 6 Temperature of refractory reactor wall at different quantity of processed Al powder (I_{arc}=250 A)

Experiments were performed in the following sequence:

- the arc was ignited the operating current to 250 A was set up. The reactor operated in "idle regime" for 3 min, which was enough time to heat the internal surface of the reactor wall to a temperature higher than 800°C (figure 6) (heating was controlled by thermocouple 11 (figure 5));

- the arc passed from regime **„rotating arc column"** into regime **„plasma volume".** The moment of that transition was found as corresponding to the change of the arc sound and to a slight voltage drop;

- charge feeding through the hollow graphitic cathode and into the reactor volume started via the powder-feeding system "PLASMALAB";

- feeded particles were taken by the plasma volume where they melted and were transported to the wall of the tube reactor. Skull was formed (figure 5) together with metal-slag melt which started flowing towards the reactor exit and entered the graphitic special design crucible placed in the tundishe. Since the water cooled power supplying graphitic bush (figure 5) was mounted in the lower section of the reactor, increase of the thickness of the skull layer 5 was observed. This cooling effect yielded a characteristic profile of the reactor longitudinal cross section (figure 7);

- the experiment proceeded for 15 more minutes after the reactor wall attained a constant temperature. This time was enough to reliably estimate how much charge was melted and landed on the reactor wall and how much of it fell non-processed into the crucible below the reactor.

- after the experiment conclusion, a profile of the tube reactor-anode was taken. The quantity of charge having melted and flown from the reactor into the crucible, as well as that of non-processed charge, were collected and weighted.

3. Results

For further technological development of the FFP reactor design, as well as for the extension of its application area, it is very important to find out what will be the effect of the described "plasma arc" metamorphosis in the "plasma volume" on the efficient transfer of powdered charge from the outlet of the hollow graphite cathode to the wall of the tubular reactor anode.

The experiments for reduction processing of the powder charge were performed under current constant value 250 A, with three different rates of charge feeding into the reactor working space – 10, 30 and 50 g/min.

Operating voltage varied from 50-55 V at arc ignition to 60-70 V when different charge quantities were fed. Reactor electric power varied within ranges 12.00 – 17.25 kW.

The experimental results are shown in figure 6 and figure 7. Significant difference in the degree of charge processing* was observed for consumption of 10 and 30 g/min (figure 6). While for minimal

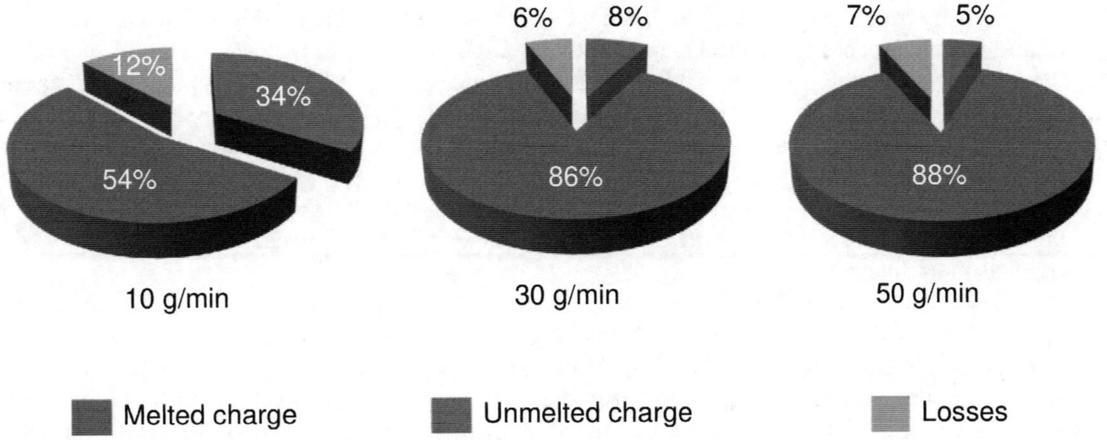

Figure 7 Efficiency of the melting of powder oxide charge in reactor "PLASMALAB" with Auto-Electro-Magnetic Rotating arc, considering different charge consumption

consumption of 10 g/min the quantity of processes charge was hardly 54%, for much larger consumption - 30 and 50 g/min, the quantity of processed charge grew up to 86 and 88%, respectively. That fact being illogical at first sight, can be explained with the arc burning mechanism, i.e. different mode of arc burning provided different charge transportation.

Charge quantity of 10 g/min was insufficient to cool the arc volume and to reverse arc burning regime from **"plasma volume"** to **"rotating arc column"**. Considering the latter, it is probable that the density of negatively charged particles moving from the cathode to the wall of the tube reactor is significantly larger. This is the probable reason for the significantly better transport capabilities of the **"rotating arc column"** as compared to those of the **"plasma volume".** That supposition is backed up by the fact that for charge consumption of 10 g/min, the skull layer is quite thin (figure 7), melt accumulated in the crucible is homogeneous and the degree of reduction of the metal oxides is low – 61-66%, i.e. metal-slag melt has been strongly overheated, it has low viscosity and it has flown down into the crucible for a short time. The insufficient time is also the reason for low reduction degree. Besides, charge non-processed particles melted when moving along the reactor axis through the plasma volume and then entered the central section of the special crucible. Large part of those non-processed oxide particles were spheroidized. This fact confirms the extremely high temperature of the "plasma volume".

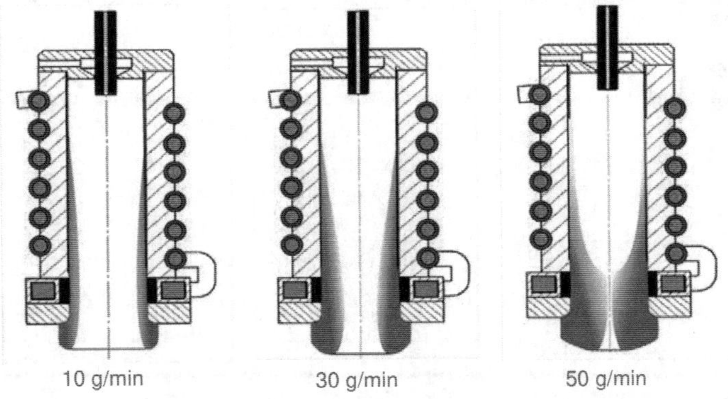

Figure 8 Scheme of a vertical cross section and profile of skull layer of "PLASMALAB" reactor with Auto-Electro-Magnetic Rotation of the arc, considering different charge consumption

---

* Consider material fed into the reactor. We adopt the term "processed charge", meaning material which the arc has transported in melted state to the reactor wall and which has flown down the wall as a thin film. Then, the partially reduced metal-slag melt accumulates in the external ring-shaped section of the graphite crucible 10 (figure 5).

As seen in figure 6, charge losses for 10 g/min consumption are the greatest ones. This is so, since evaporation of the processed material is most intensive in this case.

The increase of charge quantity yields decrease of the temperature of the "plasma volume" and it is transformed into a "rotating arc column". This is seen in the schemes of the cross sections cut along the reactor height (figure 7). Significant difference between the reactor profile for charge consumption of 10 g/min and that for charge consumption of 30 g/min is seen.

The lower reactor temperature for charge consumption of 30 and 50 g/min has led to the accumulation of a skull with significant thickness, especially in the reactor lower section where cooling was more intensive.

Consider significant increase of the quantity of charge fed for processing (over 45-50 g/min for the specific experiments discussed). Then, a tendency of extreme skull increase and a risk for reactor block up occur. Besides, increase of skull thickness leads to decrease of the reaction area of the melted charge layer where the reduction processes run most intensively. This effect is confirmed by the fact that for charge consumption of 50 g/min, the degree of reduction is 78-81 %, while for charge consumption of 30 g/min it is 88-91%.

4. Nano-materials production

The results obtained gave us confidence that the PLASMALAB FFP-reactor, after proper redesign, could be successfully used to produce powdered nano-sized materials. For the purpose, the anode design was changed and the diameter of the water-cooled tubular FFP-reactor was increased (figure 9). The anode (figure 9) was shaped as a water-cooled ring and made of copper tube. It was mounted concentrically to the cylindrical reactor, in its lower part and was electrically insulated from it.

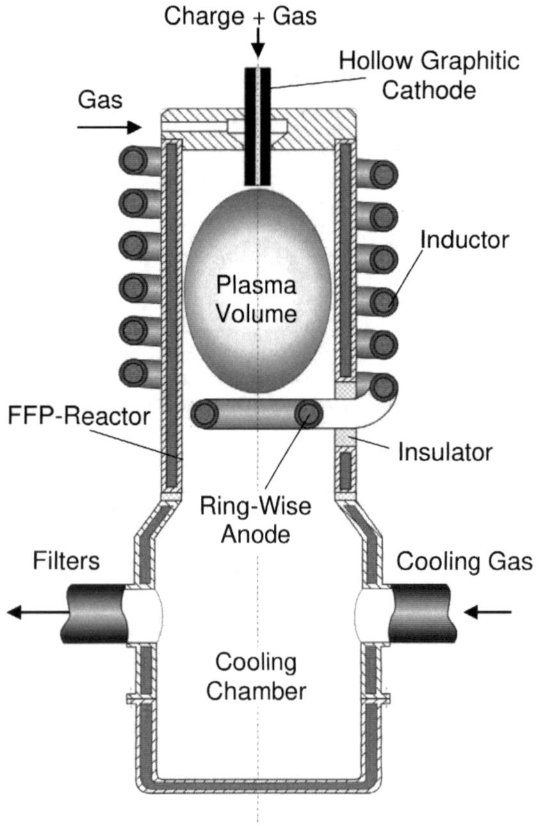

Figure 9 Schema of PLASMALAB FFP-reactor with hollow graphitic cathode and electromagnetic control of arc for nano-materials synthesize

The principle of device operation is as follows. The arc is subjected to **A**uto-**E**lectro-**M**agnetic **R**otation by the inductor serially connected to the anode circuit (figure 9). The disperse material subject to processing and the technological gas (if required) are fed through the hollow graphite cathode by the powder feeding system (figure 1).

The energy parameters of the arc in the reactor and the quantity of processed charge are kept in proportions so as to maintain a „plasma volume" combustion mode. That mode permits achievement of high average mass temperature (over 2000^0C), and the dispersed processed material passes along the reactor axis, through the plasma space where the main physico-chemical processes were running. The product leaves the reactor through the annular anode, after which it enters a cooling chamber where the obtained nano structure is fixed.

Figure 10 SEM photograph of synthesized AlN powders collected in the collection vessel

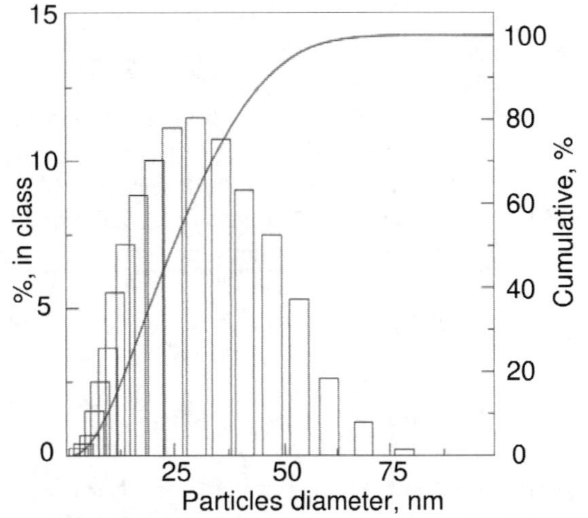

Figure 11 Size distribution of the AlN powder measured with a particle size analyzer

The reactor described was used in experiments for the synthesis of AlN from aluminum powder and nitrogen (figure 10). During each experiment 100 g Al powder was processed. An average yield of 76% was achieved. The non-reacted aluminum powder was 22%. 78% of the produced AlN with average particle size 35 nanometers (figure 11) and average specific surface 39 m^2/g separated in the cooling chamber. About 6% deposited on the reactor walls and the anode, and the rest of that quantity left the reactor with the off gases. The total charge loss with the off gases was about 16%.

5. Conclusions

- The first experiments demonstrated that the new design of PLASMALAB FFP-reactor is suitable for the production of nano-size powdered materials.
- The geometrical dimensions of the working space of reactor and the anode will be optimized in order to improve the technical and economic parameters of the process.

References

[1] Introduction to nanotechnology, Henrik Bruus, MIC – Department of Micro and Nanotechnology, Technical University of Denmark, Lyngby, spring 2004

[2] Handbook of Nanoscience, Engineering, and Technology, Second Edition, William A. Goddard Iii , Donald W. Brenner , Sergey E. Lyshevski , and Gerald J. Iafrate ,CRC Press 2007, Print ISBN: 978-0-8493-7563-7 ,eBook ISBN: 978-1-4200-0784-8

[3] Nanomaterials Handbook, Yury Gogotsi , CRC Press 2006 , Print ISBN: 978-0-8493-2308-9, eBook ISBN: 978-1-4200-0401-4

[4] MacRae D. R., Plasma Reduction of Iron Ores to Raw Steel, Presented at the Office of Technology Assessment Seminar on New Techniques in Steelmaking, May 2-3, 1979, Washington.

[5] Gold.R.G., Sandall.W.R., Cheplick,P.G. et al. Plasma Reduction of Iron Oxide with Hydrogen and Nature Gas at 100 kW and one Megawatt, Homer Research Lab., Bethlehem Steel Co., Bethlehem, PA, 1980 USA.

[6] Mihovsky M., Hadzhiyski V., Todorov L., Electromagnetic and Gas Dynamic Control of Transferred Plasma Arc in Metallurgical Plasma Reactors and Furnaces, High Temp. Material Processes 11 (2007), pp. 359-369.

[7] Mihovsky M., Hadzhiyski V., Todorov L., New Feeding System for Plasma Reactor "PlASMALAB" for Reduction Processing of Disperse Raw and Waste Materials, High Temp.

Material Processes 11 (2007), pp. 505-513.
[8] Hadzhiyski V., Mihovsky M., Speed Determination of Auto-Electro-Magnetic Rotation (AEMR) of DC Arc in "PLASMALAB" FFP-plasma Reactor, HTPP-10, Patra, Greece.
[9] M. Mihovsky, V. Hadzhiyski, Transport and Heat Phenomena of Transferred DC Arc Under Auto-Electro-Magnetic Rotation (AEMR) in PLASMALAB Reactor, BegellHouse, High Temperature Material Processes, 2009, Volume 13, Issue 2, pp. 121-135.
[10] Mihovsky M., Tzonev Tz., Advenced Plasma Furnace and Technology for Reduction Processing, 1st International Congress of Science and Technology of Ironmaking, ISIJ, Sendai, Japan, 1994, pp. 621-626.
[11] Mihovsky.M.K. et al. Cathode Unit with a Hollow Graphite Electrode for a Metallurgic Arc Plasmotron, Metallurgia, 1-2,1994.
[12] Mihovsky M., Tzonev Tz., Lucheva B., Plasma Torch-Reactor for Reduction Processing in Metallurgy, Progress in Plasma Processing of Materials, 1997, Begell House, New York, pp. 543-550.
[13] Mihovsky M., Tzonev Tz., Lucheva B., Motion Organization of Plasma Arc in Plasma Torch-FFP-Reactor System, Junior Euromat'96, August 26-30, 1996, Switzerland, pp. 436-437.

Journal of Physics: Conference Series **275** (2011) 012006 doi:10.1088/1742-6596/275/1/012006

Investigation of parameters of the three phase high-voltage alternating current plasma generator with power up to 100 kW working on steam

Ph G Rutberg, S A Lukyanov, A A Kiselev, S A Kuschev, Gh V Nakonechny, A V Nikonov, S D Popov[1], E O Serba, V A Spodobin and A V Surov

Institute for Electrophysics and Electric Power Russian Academy of Sciences, 18 Dvortsovaya emb., St.-Petersburg, 191186, Russia

[1]E-mail: sergey_popov1974@mail.ru

Abstract. The paper presents the results of experimental investigation of parameters of the three-phase high voltage alternating current plasma generator with power up to 100 kW operating on steam with gas protection of the electrodes. Researches were carried out over a range of arc current from 25 to 50 A and range of steam consumption of 3-5 g/s. Current-voltage and volt consumable characteristics, operation oscillograms and dependence of power versus the flow rate of steam and protective gas are presented.

1. Introduction

Steam-and-water plasma consists of hydrogen and oxygen acting as active reagents in many chemical reactions. There is no ballast like nitrogen is in air plasma. For many processes nitrogen is inert and, moreover, is harmful, as it is a source of toxic NOx.

According to forecasts, the future of power engineering during, at least, XXI century will be connected with use of various types of organic fuel of wood, municipal waste and mainly coal which stocks are great enough. However, significant emissions of CO2, oxides of sulfur and nitrogen of coal-consuming plants, huge scales of slag yield and non-cultivated grounds demand basic revision of methods of further use of coal as a fuel. Recently in the world the concept of complex processing of coal with production of high-calorific synthesis gas (mixture CO and H2) [1-6] and accompanying saleable material (from slag) is examined: aluminum, iron carbide, etc. The basis of this strategy is the process of coal gasification. In the future coal gasification is the most significant the practical application of steam-and-water plasma. A basic element of systems of plasma gasification and pyrolysis of coal and various organic containing waste are electric arc plasma generators. As it has been noted above, for many of these processes steam-and-water plasma is the optimum environment. Therefore problems of development and research of electric arc generators of steam-and-water plasma are rather actual.

Plasma torches generating oxygen-hydrogen plasma with temperature from 2000 up to 5000 K are necessary for technologies of plasma production of liquid synthetic fuels from coals, slates or peat. The designs using water or steam as a plasma forming environments are related to such type of plasma generators.

As a result of the researches carried out in this area by different collectives, various designs of electric arc plasma generators using water [7 - 14] or steam [15 - 20] as a plasma forming environment have been created.

The review of existing designs of electric arc plasma generators intended for generation of low temperature (up to 10000) plasma has shown the following:

Basically the presented designs are direct current plasma generators with steam supply and protection of one of electrodes (cathode or anode) by inert gas (mainly argon). A part of designs represent highly enthalpy direct current plasma generators with argon arc squeezing by a water vortex and arc closing on the external rotating electrode.

Researches have shown the increased erosion of electrodes in electric arc plasma generators, using water as a plasma forming environment. For this reason at creation of industrial steam-and-water electric arc alternating current plasma generators it is necessary to apply gas protection of electrodes. Thus the most promising protective gas environment is air, owing to such qualities as cheapness and abundance.

In the majority of designs the arc voltage drop does not exceed 300 - 400 V. That is why the increase in power is reached due to increase in arc current that negatively affects on operating time of electrodes.

As steam-and-water plasma represents a mixture of hydrogen plasma with oxygen in the ratio 2:1, its properties are closer to properties of hydrogen as in its content in the mixture is higher. Heat conductivity of steam-and-water plasma is significantly higher, and viscosity considerably lower, than at other kinds of plasma. Owing to this at arc burning in a vortical stream of steam-and-water plasma the arc column is subjected to not so strong lateral cooling, as in the environment of pure hydrogen, but to much more intensive, than in the environment of other gases. It, in a combination with a high specific thermal capacity of steam-and-water plasma, leads to greater intensity of the electric field in the arc column and to greater enthalpy of steam-and-water plasma at the same arc current that imposes essential requirements of the power supply system and constructional features of the electric discharge chamber.

Based on the carried out preliminary researches the decision to develop a design of a resource electric arc plasma generator (200 hours of continuous work) for generation of water plasma using the principles incorporated in designs of high-voltage electric arc alternating current plasma generators [21] - increase in power of plasma generator due to increase in arc voltage was made.

2. Description of the experimental plant

The power supply for single-phase or three-phase load comprising the electric arc has been constructed for the tests. The schematic of the power supply source is shown in Figure 1.

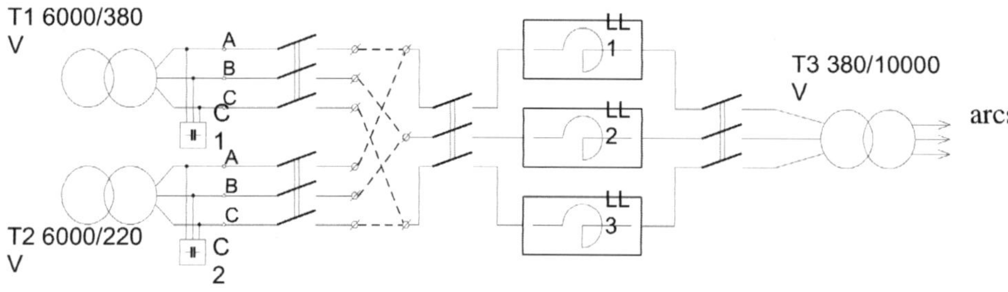

Figure 1. The electrical schematic.

The power supply system is assembled on the basis of a step-up transformer 380/10000 V. Inductances LL are parallel or (serial) connection of the reactors P1500 and P1800. Switching of the output voltage is realized by choice of the supply network: 220 V or 380 V. Characteristics of the

Journal of Physics: Conference Series **275** (2011) 012006 doi:10.1088/1742-6596/275/1/012006

power source: no load voltage is 6000 or 10000 V, short circuit current 30, 40 and 50 A in dependence on the scheme of the reactor connection.

The experimental design of the electric arc plasma generator has been created for carrying out tests. This design has an electric arc chamber optimized for operation with steam at temperature of 250°C. It allows supply into the electric-discharge chamber of various plasma forming media. Figure 2. shows the external view of the experimental plasma generator.

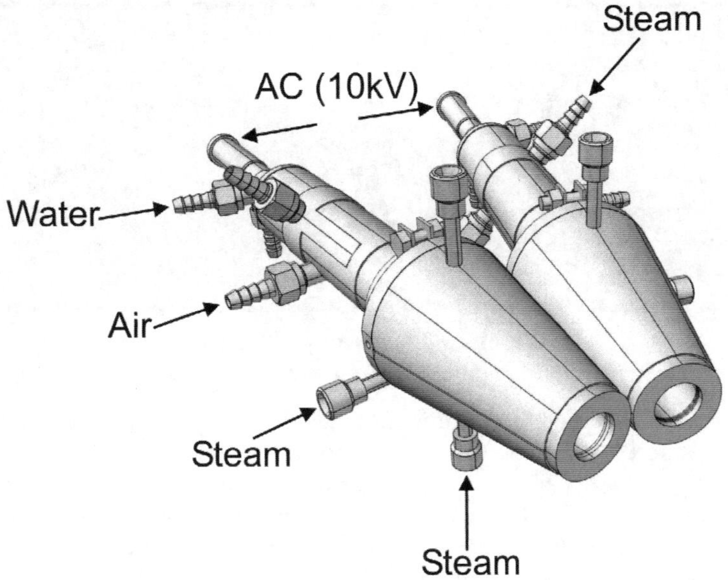

Figure 2. Experimental steam-air high-voltage alternating current plasma generator.

The experimental plasma generator consists of chambers for protective gas supply, steam, nozzle apparatus (conicity 40%), electrode system of water-cooled rod electrode (water pressure in the cooling system is 1.2 MPa). Electrodes with 8 mm diameter are made of a ceramic-metal alloy. Steam pressure can vary from 0.2 up to 1 MPa, temperature is 200°C. Specific wear of electrode material is ~10^{-6} g/C. Electrode lifetime is more than 200 hours. Plasma generator efficiency measured on thermal balance (W_{arc}/W_{cons}) is about 0.75-0.92 depending on parameters of gas supply.

3. Experiment results

A series of experiments, where the following parameters have been varied, are carried out: characteristics of the power supply (current); steam flow rate; air flow rate; proportion of a steam-air mixture. Steam flow rate varied from 1 up to 4 gram per second. The range of change of the steam and protective gas flow rate for this model of the plasma generator was selected from a condition of steady arc burning. The protective gas (air) flow rate varied over the range from 0.5 up to 5 g/s. The short circuit current of the power supply varied from 12 up to 33 A. Figure 3. shows the photo of plasma jet of the experimental plasma generator.

Figure 3. Steam-air plasma jet

Arc intensity E ~ 50 – 100 V/cm depending on mode of supply of plasma and protective gas. The dependence of voltage arc drop variation from component composition and mass flow rate of plasma gas was investigated during experiments. For the investigated high-voltage steam-air alternating current plasma generator the arc voltage drop depends not only on the total flow rate of plasma environments, but also from a relation of protective air and steam flow rate. Figure 4 shows the character of variation of voltage drop on the plasma generator arc versus the flow rate of the mixture of plasma gases.

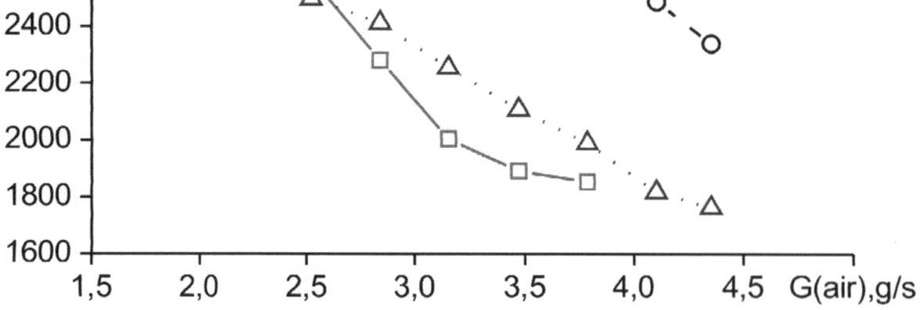

Figure 4. Volt-flow rate characteristics of the single-phase electric-arc plasma generator operating on steam-air mixture.

Researches have shown that dependence of voltage variation on the mass flow rate of steam-air mixture essentially differs from results obtained at use of air in plasma generators of similar design [21,22,23] or carbonic gas, where with increase in the mass flow rate of plasma gas it is observed growth of arc voltage drop in connection with compression of arc column by gas stream. For air-steam mixture the tendency of change of arc voltage drop is different. With increase in total air flow rate from minimal at constant steam flow rate, arc voltage drops and only at attainment of the air flow rate comparable with the steam flow rate it is observed the tendency of growth of arc voltage drop (similarly to high-voltage electric arc alternating current plasma generators working only on air).

It is necessary to note, that the arc current practically does not change over all range of flow rates (it is stabilized by the power supply system), the current density remains constant as the cross-section of the arc column is limited by the nozzle device. Hence, for change of the arc voltage drop (arc intensity) is responsible the conductivity of the arc column. As conductivity in these conditions is defined by a relation of sections of electron-nuclear and electron-ionic interactions and at achievement of critical concentration of electrons, when section of electron-ionic collisions becomes determining, conductivity practically does not change, as a rule, at electron concentration ≥ 10^{15} cm $^{-3}$ [24]. In the examined case, considering complex plasma composition, these relations are varied. And the task of the further researches is definition of critical concentration for a concrete mixture of plasma forming environments, that, apparently, determined the character of dependence of intensity versus the flow rate.

On the basis of the obtained results the multi-phase high-voltage alternating current plasma generator was designed and created. It is intended for operation on mixture of steam and air or carbon dioxide. The external view of the plasma generator and photo of the plasma generator operation on the steam-air mixture are shown in Figure 5.

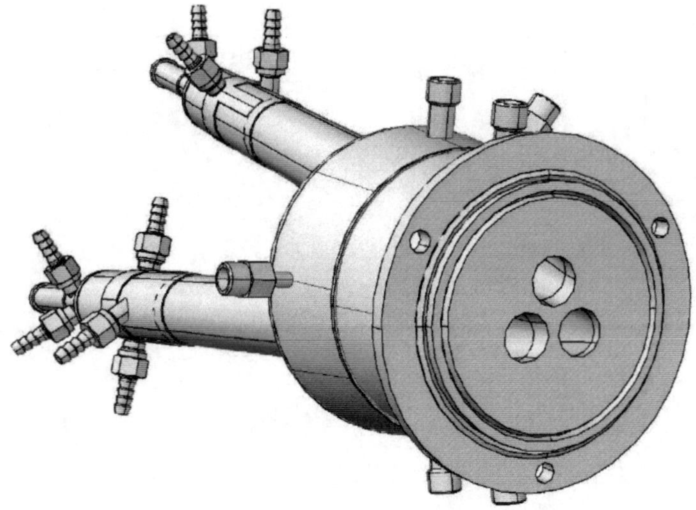

Figure 5. The experimental industrial steam-air three-phase high-voltage AC plasma generator. External view

Figure 6 and Figure 7 show the characteristics of this plasma generator working on a steam-air mixture.

Journal of Physics: Conference Series **275** (2011) 012006 doi:10.1088/1742-6596/275/1/012006

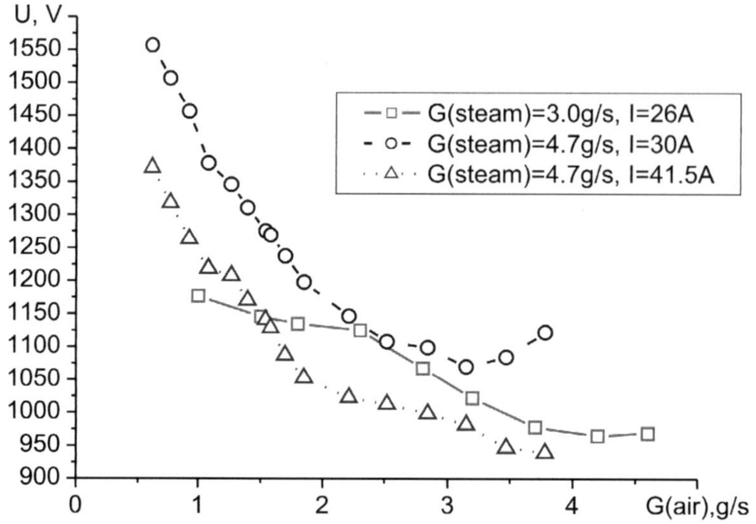

Figure 6. Volt-flow rate characteristics of the three-phase electric-arc plasma generator operating on steam-air mixture.

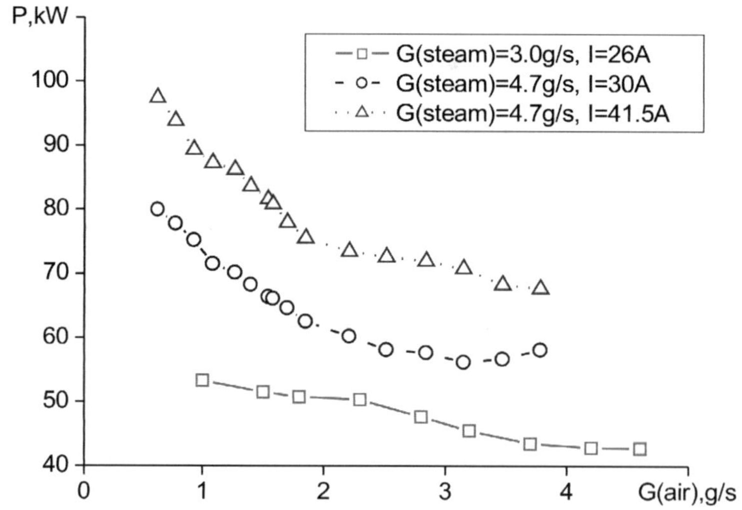

Figure 7. Dependence of power of the three-phase electric arc steam-air plasma generator versus air flow rate.

As is seen from the presented graphs the highest arc voltage drop and the highest power are achieved at the minimal air flow rate.

The efficiency of three-phase electric-arc steam-air plasma generator at relation of flow rates 3 g/s steam, 1 g/s air power 68 kW is 93%.

Photos of the experimental start up of this model of the plasma generator operating on mixture of steam and carbon dioxide are shown in Figure 8. The experimental results are presented in Figure 9 and Figure 10.

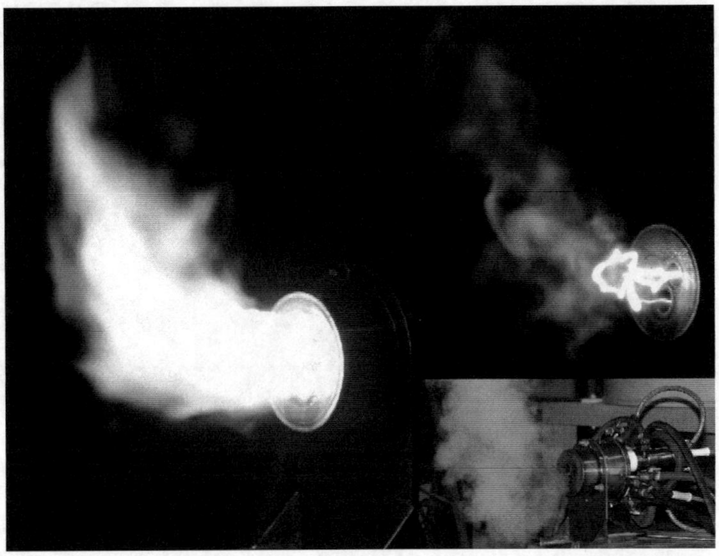

Figure 8. Three-phase high voltage AC plasma generator operation on the mixture of steam and carbonic gas. At the left and on the right above - photos of a flame and arcs of the working plasma generator accordingly; on the right below - a photo of the switched-off plasma generator with supplied steam.

Figure 9 shows the dependences of variation of the arc voltage drop versus carbon dioxide flow rate at constant flow rate of steam.

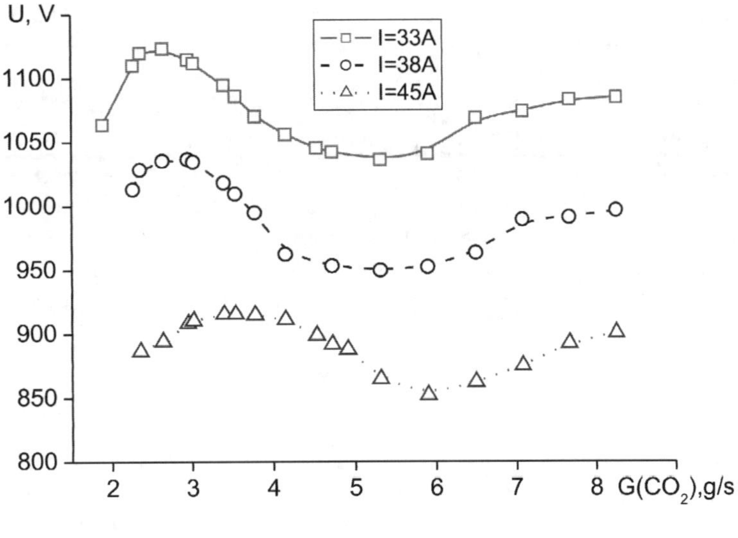

Figure 9. Volt-flow rate characteristics of the three-phase electric-arc plasma generator operating on steam-carbon dioxide mixture. Steam flow rate is 3 g/sec.

Figure 10 shows the dependences of variation of power of the electric arc multiphase high-voltage plasma generator operating on mixture of steam with carbon dioxide.

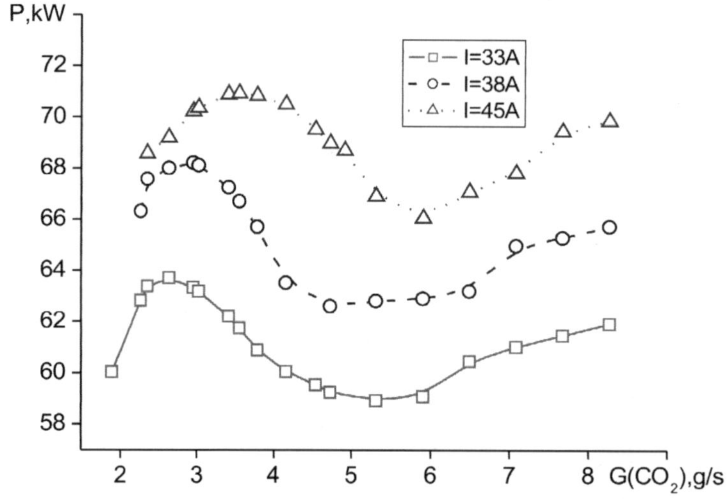

Figure 10. Dependence of power of the three-phase electric-arc plasma generator versus the parameters of the power source and flow rate of the carbon dioxide. Steam flow rate is 3g/sec.

The obtained values of variation of power versus the relation of consumption parameters of plasma agents and parameters of the power source allow the use of the created model of the plasma generator in the pilot plasmachemical reactor.

4. Conclusion

As a result of the carried out work the complicated dependence of ranges of steady-state work of the plasma generator, arc voltage drop and power from parameters of the power supply and flow rates of plasma gases have been obtained.

Values of arc voltage drop are achieved at work on a steam-air mixture 3400 V for single-phase experimental installation and 1550 V for the multiphase industrial pilot device.

The design of an experimental multiphase AC plasma generator working on a mixture of steam with air or carbon dioxide as a plasma gas has been created for the realization of the processes of plasmachemical processing of hydrocarbonic raw material.

Experiments have shown the existence of complicated dependence between the mass flow rate of steam-gas mixture, relation of steam to the second gas and electric arc voltage drop for the given configuration of the plasma generator.

Acknowlegements

The work is supported by RFBR grants 09-08-00939-a, 09-08-00926-a, 08-08-00435-a, 09-024-01087-a, 08-08-00571-a and 08-08-90413-Ukr_a

References

[1] Kuznetsov V A, Bratsev A N, Kovshechnikov V B, Kumkova I I, Popov V E, Shtengel S V et al. 2007 Distinctive features of biomass gasification using ac plasma generators working on air 2007 *IEEE Pulsed Power Conference (PPPS-2007). Digest of Technical Papers 1976-2007.*

IEEE Catalog Number: 07CH37864C. ISBN: 1-4244-0914-4. DVD-ROM. Omnipress, Madison WI USA **1223-1226**.

[2] Rutberg Ph G, Tendler M, Van Oost G, 2005 Plasma based waste treatment and energy production *Plasma Physics and Controlled Fusion, Institute of Physics and IOP Publishing Limited, Bristol* **47** A219-A230 (http://stacks.iop.org/0741-3335/47/A219)

[3] Prospects of production of superpure liquid fuels from coal 2006 *Neftegazovie technologii 2* 54 (in Russian)

[4] Amouroux J, Morvan D, Cavadias S et al 2005 The control of environmental contamination and processes of cleaning by plasma methods *Jurnal Technicheskoi Fiziki.* **75 Issue 5** 73-82 (in Russian)

[5] Rutberg Ph G 2003 Plasma Pyrolysis of Toxic Waste *Plasma Physics and Controlled Fusion* **45** 957-969.

[6] Hrabovsky M, Konrad M, Kopecky Vet al. 2006 Pyrolysis of wood in arc plasma for syngas production *High Temperature Material Processes* **10 -4**. 557-570

[7] US patent 3,649,497

[8] US patent 3,658,673

[9] US patent 3,712,996

[10] US patent 3,866,089

[11] Hrabovský M 1998 *Pure & Appl. Chem.,* **70 -6.** 1157-1162

[12] Kavka T, Chumak O, Sember V, Hrabovský M, 2007 *28th ICPIG 2007 Proceedings ISBN 978-80-87026-01-4 Published by Institute of Plasma Physics AS CR v.v.i.* 1819-1822

[13] Hrabovský M, 1997 Water-stabilized plasma generators, *Proc. 13 th Inter. Symp. On Plasma Chemistry, Beijing China* **1** 145-148

[14] US patent 2,906,858

[15] Glocker B, Borck V and Nentwig G 1999 Steam Operated Small Plasma Torches *Proc. 14th Inter. Symp. On Plasma Chemistry* 2037-2042

[16] Glocker B, Nentwig G, Messerschmid E 1- 40 kW Steam Respectively Multi-Gas Thermal Plasma Torch *Vacuum* **59** 35-46

[17] Romualdas K, Romualdas J and Vladas M Water vapour Plasma Torch for Hazardous Waste Treatment *Annals of the New York Academy of Sciences* **891** 43 – 48

[18] Watanabe T, 2005 Water Plasma Generation under Atmospheric Pressure for Waste Treatment *ASEAN Journal of Chemical Engineering* **5 - 1** 30-34

[19] Knak A N, Khvedchin I V, Ermolaeva E M, Mosse A L 2000 Characteristics of steam-plasma-generator installation with use of plasmatron PDS-3 *Transactions of IV Minsk International Forum on Heat and Mass Exchang* p. 35-39 Minsk (in Russian)

[20] Predtechenskii M R, Tukhto O M, 2006 Plasma Torch with Liquid Metal Electrodes *High Energy Chemistry* **40 -2** 119–124

[21] Rutberg Ph G, Popov S D, Safronov A A, Serba E O and Nakonechny Gh V 2007 High-Voltage Alternating Current Plasma Generators with Power up to 50 kW for Plasmochemical Applications *ICPIG 2007 Proceedings ISBN 978-80-87026-01-4 Published by Institute of Plasma Physics AS CR v.v.i.* 1762-1765

[22] Rutberg Ph G, Gorbunov V P, Kuschev S A, Lukyanov S A,.Nakonechny G V, Popov Sergey D et al 2007 Characteristic Features of operation of High-Voltage Electric Arc Plasma Generators with Rod Electrodes and power from 5 up to 50 kW in a Pilot Plasmachemical Unit *IEEE Catalog Number 07CH37865, ISBN 1-4224-0914-4, Library of Congress 81-644315* 603

[23] Rutberg Ph G, Safronov A A, Surov A V and Popov S D 2007 Alternating Current Electric Arc Plasma Generators *ICPIG 2007 Proceedings ISBN 978-80-87026-01-4 Published by Institute of Plasma Physics AS CR v.v.i.* 1774-1777

[24] Rutberg Ph G 1970 *Some questions of investigation of gas discharge plasma and creation of high magnetic fields-* (L.:Nauka.Lenigr.department, in Russian)

Journal of Physics: Conference Series **275** (2011) 012007 doi:10.1088/1742-6596/275/1/012007

The role of radiation losses in high-pressure blasted electrical arcs

J Gregor[1], I Jakubova[2,4], J Senk[1] and A Maslani[3]

[1] Brno University of Technology, Faculty of Electrical Engineering and Communication, Department of Electrical Power Engineering, Technicka 8, 61600 Brno, Czech Republic
[2] Brno University of Technology, Faculty of Electrical Engineering and Communication, Department of Radio Engineering, Purkynova 118, 61200 Brno, Czech Republic
[3] Institute of Plasma Physics, Academy of Sciences of the Czech Republic, Za Slovankou 3, 182 00 Prague, Czech Republic

E-mail: jakubova@feec.vutbr.cz

Abstract. The paper deals with experiments carried out on an arc heater where the electric arc is stabilised by flowing working gas. Measured quantities (especially arc current, voltage drop, gas flow rate, and energy loss) serve as input data for a mathematical model of the arc inside a cylindrical anode channel. Previously, the losses of cathode and anode spots were assumed to be negligible in comparison with the total loss. In the new sets of experiments, a modular structure of the arc heater has made it possible to separate the losses of anode and cathode from the energy losses of the arc itself. Furthermore, the losses caused by radial conduction have been introduced into the model. The most significant change of the model concerns the computation of radiation losses of the arc. In the original model, radiation losses were taken as a portion ε of the total input power. In the modified model, the radiation loss is expressed using a theoretically calculated net emission coefficient of argon (by V. Aubrecht and M. Bartlova). This approach is possible due to a more precise determination of the arc net energy loss which results in flatter radial temperature profiles. Axial distribution of energy loss for the original and modified model is given in figures.

1. Introduction

Transport and thermodynamic properties of the working gas and the mechanisms of energy exchange between electric arc and its surroundings play a substantial role in the design and operation of various devices with intensively blasted electric arcs, such as arc heaters and circuit breakers. External integral quantities characterising the device, especially the current, voltage, gas flow rate etc., can be sometimes measured but the internal processes often remain hidden. Mathematical models can exploit measured data for discovering and describing the phenomena running inside.

Many contributions have dealt with switching electrical arcs. With a different stage of simplification the behaviour of the electric arc burning in flowing gas has been analysed. For instance

[4] I Jakubova, Brno University of Technology, Purkynova 118, 61200 Brno, Czech Republic

Journal of Physics: Conference Series **275** (2011) 012007 doi:10.1088/1742-6596/275/1/012007

Frost and Liebermann [1] presented computed properties of hexafluoride and a simplified model of enthalpy flow in high-current interval. Ragaller [2] used data obtained from experiments with a quasi-stationary, high-pressure axially blasted electrical arc of cylindrical shape for modelling. Approximation of radial temperature dependencies with a generalized parabola in the arc and with a Gaussian function outside the arc recommended by Ragaller seems to be suitable also for our experiments and modelling. Lowke, Ludwig and Tuma [3], [4] dealt with a simple model of stabilised electric arcs. Lowke, Stokes and Ramakrishnan [5] designed one-dimensional models of free high-current electric arc with the influence of turbulence included by means of Prandtl mixing length. Their relations between the arc cross-section and current were used in the presented model with the exponent n_r as one of the computational parameters.

The authors try to design a simple model of electric arcs stabilized by a flowing gas in a cylindrical arc heater channel. The model should exhibit no special requests for software and hardware but it should clearly describe especially the energy exchange between the arc and its surroundings. As mentioned above, some relations of older work can be utilized. Previous computations with an older version of the model indicated a prevailing role of the arc radiation in the total power loss of the arc. This fact turns attention to a more adequate expression of the radiation term in the energy balance. For this purpose, the net emission coefficient of argon [6], [7], [8] has been used with a necessary modification caused by the non-ideal real experimental situation.

This paper is organised as follows: Section 2 describes the mathematical model with special attention to the newly introduced modifications. Section 3 deals with the experimental device that served as a source of the input experimental data. Section 4 compares the results of three comparable experiments and computations and demonstrates the influence of the designed modifications of the model. Section 5 concludes the paper.

2. Mathematical model of an electric arc: Basics and advanced modifications

The basics of the model of the electric arc designed by the authors are explained in detail in [9], [10]. Here, only necessary basic relations are mentioned while attention is focussed on advanced modifications. The aim is to get a more precise description of the physical phenomena inside the anode channel.

The model is based on the mass and energy conservation equations and Ohm's law. The arc plasma is supposed to be in local thermodynamic equilibrium. It is assumed the kinetic energy is small compared to the enthalpy. Presumed cylindrical symmetry of the arc follows from the shape of the anode channel. A further simplifying presumption is that the Mach number M is constant over the channel cross-section. In the energy balance of individual elementary cylinders ΔV_A and cylindrical rings ΔV_C of the channel (see figure 1), conductive and radiation energy flow in the radial direction and enthalpy flow in the axial direction are considered as highly prevailing components.

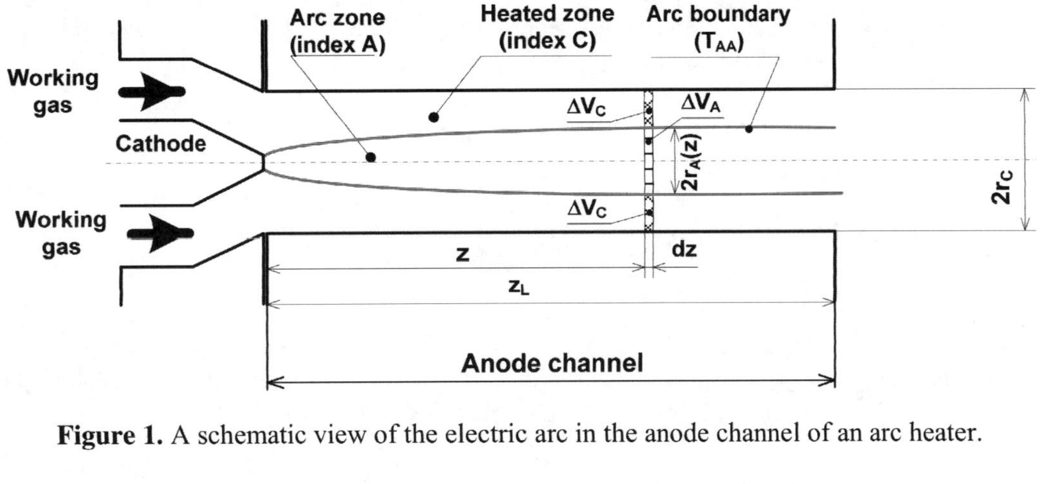

Figure 1. A schematic view of the electric arc in the anode channel of an arc heater.

A schematic view of the electric arc in the anode channel of an arc heater is given in figure 1. The anode channel of the radius r_C and the length z_L is divided into a cylindrical arc zone and a remaining ring volume. The arc zone is characterised by high plasma conductivity. On the arc zone boundary, plasma conductivity sharply falls nearly to zero. The corresponding temperature T_{AA} at the arc boundary depends on the used working gas ($T_{AA} = 6000$ K for argon). The quantities in the arc zone are indicated with subscript A. In the remaining ring volume between the arc zone and the channel wall the temperature is bellow T_{AA} and can be as low as the temperature of the cold working gas at the input T_0. In the region of relatively low temperatures $T < T_{AA}$ some temperature dependencies of transport and thermodynamic properties of the working gas can be expressed in simple forms, which makes the computation simpler and faster. Subscript C is used to indicate quantities and functions in this zone.

The axial development of the arc is described by the dependency of the arc boundary radius $r_A(z)$ [5]

$$r_A(z) = r_0\left[1 + \left(\frac{z}{r_0}\right)^{1/n_r}\right] \tag{1}$$

where r_0 is the radius of the arc at the cathode, defined by the arc current and by the cathode current density (10^8 Am^{-2} for currents up to 2.16 kA), while n_r is a calculated parameter.

Another calculated parameter is the exponent n in the relation describing the radial dependency of temperature in the arc zone

$$T_A(r,z) = T_0 + (T_{AA} - T_0)\left\{\left[\left(\frac{T_A(0,z) - T_0}{T_{AA} - T_0}\right)^{1/n} - 1\right]\left[1 - \frac{r^2}{r_A^2(z)}\right] + 1\right\}^n \tag{2}$$

where T_{AA} is the temperature for which the conductivity of the working gas sharply falls to zero (6000 K for argon) while T_0 is the temperature of cold gas at the input. Outside the arc zone, the radial temperature dependency is approximated by a Gaussian function as follows

$$T_C(r,z) = T_0 + (T_{AA} - T_0)\exp\left(\frac{r_A^2(z) - r^2}{b_C^2(z)}\right). \tag{3}$$

Both function values and derivatives of (2), (3) at the arc boundary must be equal to ensure a smooth radial temperature dependency together with equal conductive heat flows at both sides. This demand gives the following expression for $b_C(z)$

$$b_C(z) = r_A^2(z)\left\{n\left[\frac{T_A(0,z) - T_0}{T_{AA} - T_0}\right]^{1/n} - 1\right\}^{-1}. \tag{4}$$

With the above mentioned presumptions, continuity and energy conservation equations can be written for the arc zone (subscript A) and the rest (heated and cold zone, subscript C) as follows

$$M(z)\left\{\int_0^{r_A(z)} \rho[T_A(r,z)]a[T_A(r,z)]2\pi r\mathrm{d}r + \int_{r_A(z)}^{r_C} \rho[T_C(r,z)]a[T_C(r,z)]2\pi r\mathrm{d}r\right\} = G, \tag{5}$$

$$M(z)\left\{\int_0^{r_A(z)} \rho[T_A(r,z)]a[T_A(r,z)]h[T_A(r,z)]2\pi r\mathrm{d}r + \int_{r_A(z)}^{r_C} \rho[T_C(r,z)]a[T_C(r,z)]h[T_C(r,z)]2\pi r\mathrm{d}r\right\} = U(z)I - P_L(z) \tag{6}$$

Here, M is Mach number, ρ the specific mass, a the sound velocity, h the enthalpy, G the gas flow rate, $U(z)$ the voltage at z axial position, I the arc current, and $P_L(z)$ the power loss. The left-hand side of the equation (6) expresses the enthalpy flow $P_h(z)$ at the axial position z.

Having approximated the radial temperature dependency (2), integration over the arc radius can be converted into integration over temperature. Generally expressed for integrating a function $\{f[T_A(r,z)]2\pi r\mathrm{d}r\}$ in the arc zone, the transform is the following

$$\int_0^{r_A(z)} f[T_A(r,z)]2\pi r\,\mathrm{d}r = \pi r_A^2(z)F[T_A(0,z),n], \tag{7}$$

$$F[T_A(0,z),n] = \left\{n\left[(T_A(0,z)-T_0)^{1/n} - (T_{AA}-T_0)^{1/n}\right]\right\}^{-1} \int_{T_{AA}}^{T_A(0,z)} f(T-T_0)(T-T_0)^{\frac{1-n}{n}}\,\mathrm{d}T. \tag{8}$$

Using (7), (8) and suitable approximations of the temperature dependencies of the specific mass ρ, sound velocity a, enthalpy h and electric σ and thermal λ conductivity, the necessary functions can be computed beforehand which makes the solution much easier. In the following text, $RA[T_A(0,z)]$ stands for integration (8) with $f(T-T_0)$ equal to the product of the specific mass $\rho(T-T_0)$ and sound velocity $a(T-T_0)$, thus $RA[T_A(0,z),n] = \left\{n\left[(T_A(0,z)-T_0)^{1/n} - (T_{AA}-T_0)^{1/n}\right]\right\}^{-1} \int_{T_{AA}}^{T_A(0,z)} \rho(T-T_0)a(T-T_0)(T-T_0)^{\frac{1-n}{n}}\,\mathrm{d}T$.

Similarly $RAH[T_A(0,z)]$ stands for integration (8) with $f(T-T_0) = \rho(T-T_0)\ a(T-T_0)\ h(T-T_0)$, thus $RAH[T_A(0,z),n] = \left\{n\left[(T_A(0,z)-T_0)^{1/n} - (T_{AA}-T_0)^{1/n}\right]\right\}^{-1} \int_{T_{AA}}^{T_A(0,z)} \rho(T-T_0)a(T-T_0)h(T-T_0)(T-T_0)^{\frac{1-n}{n}}\,\mathrm{d}T$ and

for $f(T-T_0) = \sigma(T-T_0)$ $\Sigma[T_A(0,z),n] = \left\{n\left[(T_A(0,z)-T_0)^{1/n} - (T_{AA}-T_0)^{1/n}\right]\right\}^{-1} \int_{T_{AA}}^{T_A(0,z)} \sigma(T-T_0)(T-T_0)^{\frac{1-n}{n}}\,\mathrm{d}T$.

With temperatures lower than T_{AA}, the temperature dependencies of ρ, a, h and their products can be expressed in very simple forms [10].

The axial dependency of centreline temperature $T_A(0,z)$ is calculated in individual elements of the channel volume. The total length z_L is divided into m elements Δz. The energy equation for each segment states that the increment of the input electric power $\Delta U_{ch}\ I$ is equal to the sum of the increments of the enthalpy flow ΔP_h, conductive ΔP_{con} and radiation power loss ΔP_r

$$\Delta U_{ch}(z_i)I - \Delta P_h(z_i) - \Delta P_{con}(z_i) - \Delta P_r(z_i) = 0. \tag{9}$$

The value of the Mach number in the ith element can be derived from the continuity equation and reads

$$M(z_i) = G\left\{\pi r_A^2(z_i)RA[T_A(0,z_i),n] + p\left(\frac{\kappa M_\rho}{R}\right)^{\frac{1}{2}} \int_{r_A(z_i)}^{r_c} T_C^{-\frac{1}{2}}(r,z_i)2\pi r\,dr\right\}^{-1} \tag{10}$$

where G is the gas flow rate, p the pressure, κ Poisson constant, M_ρ the molar mass, R the universal gas constant, r_c the channel radius, T_C the temperature outside the arc zone. The individual terms of (9) can be expressed as follows

- The increment of the input power of the ith element

$$\Delta U_{ch}(z_i)I = E(z_i)\Delta zI = I^2\Delta z\left\{\pi r_A^2(z_i)\Sigma[T_A(0,z_i),n]\right\}^{-1} \tag{11}$$

- The increment of the enthalpy flow of the ith element

$$\begin{aligned}\Delta P_h(z_i) = {} & \pi\left\{M(z_i)r_A^2(z_i)RAH[T_A(0,z_i),n] - M(z_{i-1})r_A^2(z_{i-1})RAH[T_A(0,z_{i-1}),n]\right\} \\ & + 2\pi pC_p\left(\frac{\kappa M_\rho}{R}\right)^{\frac{1}{2}}\left\{M(z_i)b_C^2(z_i)\left[\sqrt{T_{AA}} - \sqrt{T_C(r_C,z_i)}\right] - M(z_{i-1})b_C^2(z_{i-1})\left[\sqrt{T_{AA}} - \sqrt{T_C(r_C,z_{i-1})}\right]\right\}\end{aligned} \tag{12}$$

where C_p is the thermal capacity of the working gas at $T<T_{AA}$.

- The increment of the conductivity loss of the ith element

$$\Delta P_{\mathrm{con}} = 4\pi\Delta z\left\{r_{\mathrm{C}}^{2}\frac{\lambda[T_{\mathrm{C}}(r_{\mathrm{C}},z_{i})]}{b_{\mathrm{C}}^{2}(z_{i})}[T_{\mathrm{C}}(r_{\mathrm{C}},z_{i})-T_{0}]\right\} \tag{13}$$

where λ is the thermal conductivity. If temperature T_{C} is nearly as low as T_0, this term can be neglected.

- The increment of the radiation loss of the *i*th element

$$\Delta P_{\mathrm{r}}(z_i) = 4\pi^2\Delta z\, k_{\mathrm{r}}^2 r_{\mathrm{A}}^2(z_i)\varepsilon_{\mathrm{N}}[T_{\mathrm{A}}(0,z_i),k_{\mathrm{r}}r_{\mathrm{A}}(z_i)] \tag{14}$$

where ε_{N} is the net emission coefficient of gas which depends on the centreline temperature $T_{\mathrm{A}}(0,z)$ and the effective arc radius $k_{\mathrm{r}}r_{\mathrm{A}}(z)$ [6], [7], [8]. In the previous model [9], [10] the radiation loss of the arc segment was computed as a fixed portion of the input power $\Delta P_r(z_i) = (1-\varepsilon)\,\Delta U(z_i)I$. The parameter $\varepsilon = P_{\mathrm{L}}/UI$ was determined as a ratio of the measured total loss of the channel and the total input power. Thus the same value was taken for all segments with no respect to their temperature and radius. The net emission coefficient is defined by the power irradiated from a homogeneous gas cylinder of the radius R and the constant temperature T. In real experiments, this condition is hardly fulfilled. That is the reason for introducing the calculation parameter k_{r}. The radiation of the arc slice is computed using the net emission coefficient of the gas cylinder of the effective radius $k_{\mathrm{r}}r_{\mathrm{A}}$ and temperature $T_{\mathrm{A}}(0,z)$. In comparison with the previously applied relation, the approximate description of the radiation loss using the net emission coefficient ε_{N} and the effective arc radius $k_{\mathrm{r}}r_{\mathrm{A}}$ at least partly respects the connection of the irradiated power and the temperature and radius of arc segments.

During the computation, a set of parameters n, n_{r} and k_{r} is searched for with which measured and computed integral values of voltage drop, total power loss and output enthalpy flow differ mutually less than by accepted differences (several percent). For a chosen set of parameters n, n_{r} and k_{r} the energy equation of individual segments is solved and the centreline temperature is determined. Finally, the increments of the voltage and the radiated power loss of the segments are summed along the whole channel length and the output enthalpy flow is computed. The computed and measured values are compared and the computation is either finished or repeated with a modified set of parameters until a satisfying agreement is reached.

3. Experimental set-up

Experimental input data for the designed model of the electric arc burning in the working gas flowing through a cylindrical channel have been collected during numerous experiments. The arc heater used has been designed by the authors as a modular device for a laboratory purpose. Both the cylindrical anode channel and other possible parts (anode, expanding chamber etc.) can be compact or built of several segments. The lengths, radii, electric and cooling circuits' connections can be changed according to the aims and demands of an experiment. The dimensions of the core part of the device are in order of centimetres (e. g. the channel radius between 4 and 8 mm, the channel length up to 125 mm). The input electric power is up to 30 kW and the working gas flow rate up to a few tens of grams per second depending on the medium used (argon, nitrogen, methane, or their mixtures).

The set of measured data includes mainly basic electrical parameters (the arc current I, the voltage drop U), the flow rate of working gas G, and flow rates and input and output temperatures of cooling water going through individual segments of the arc heater.

Originally, the entire measured voltage drop U was taken as arising in the arc column and the power losses of the cathode and anode spots were supposed to be negligible. More precise measurements on the reconfigured arc heater with separated input and output segments showed that this presumption is too rough. The arc column voltage drop U_{ch} can be obtained from the total measured voltage U

$$U_{\mathrm{ch}} = U - 2\frac{P_{\mathrm{cat}}}{I}\,. \tag{15}$$

P_{cat} is the measured power loss of the cathode. The measured power loss of the output part of the anode channel includes not only the power loss of the anode spot, but also of the arc burning in this output segment. To separate them, the loss of the anode spot is taken to be equal to that of the cathode spot. Similarly, the loss of the arc P_L can be determined from the total measured loss P_{tot} of all segments

$$P_L = P_{tot} - 2P_{cat}\,. \tag{16}$$

4. Results and discussion

Three experiments discussed here as examples have been carried out under similar conditions: with the same channel radius of 8 mm, the same argon flow rate of 22.5 g/s, and the same arc current of 122 A.

The first case is referred to as the "original" experiment and evaluation. In this experiment the loss of the cathode and of the anode part of the channel were neglected and the entire voltage U= 111.8 V was taken as the voltage drop over the arc column. As explained above, the radiation of the arc column was computed as the certain portion (1-ε) of the input power with no relation to the radius and temperature of the arc. The resulting exponents were found n_r=3.2 and n=1.0.

In the second and third case, the arc heater was reconfigured in order to measure the power loss of individual parts of the arc heater. The power losses of the cathode and of the output part of the channel (including the anode part of the arc and the anode spot) were measured separately. In the second experiment ("corrected") the measured voltage after correction was lowered (15) to U_{ch}=81 V. The computation of the radiation loss was the same as in the first experiment. While the exponent n_r=3.2 remained the same as that found in the first case, the exponent n=0.1 was substantially different.

In the third case the correction of the voltage was applied as well but a longer channel was tested (z_L=108 mm instead of 80 mm). In the third case (referred to as "modified") the radiation loss was computed using the net emission coefficient of argon and the effective radius $k_r r_A$ (14). Although both exponents n_r=3.2 and n=0.1 were the same as those found in the second case, some differences between the computed dependencies can be found.

In figure 2 substantial differences between radial dependencies of temperature between the "original" case on one side and both other cases on the other side can be seen. If the losses of the cathode and anode spots are not extracted from the measured values the resulting dependencies are strongly influenced. The radial dependencies for z=78 mm are given as an example. In the first case, the computed exponent n = 1.0 leads to a rather sharp maximum of the radial temperature dependency

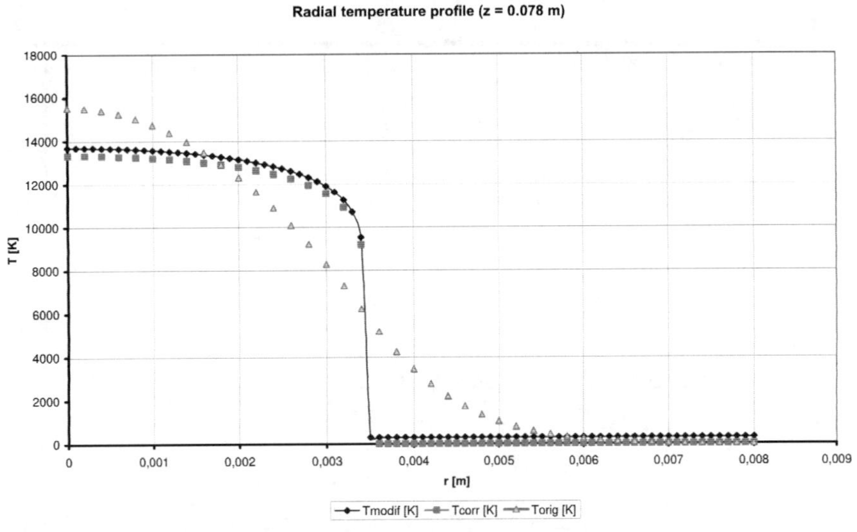

Figure 2. The radial temperature profiles for the "modified", "corrected" and "original" case.

in the arc zone. A zone of heated gas can be seen near the arc boundary. In the second ("corrected") and third ("modified") cases, the substantially lower exponent n=0.1 results in the flat radial dependency of the temperature in the arc zone. The gas outside the arc zone remains cold due to a very high derivative of the radial temperature dependency at the arc boundary (2), (3), (4).

Figure 3 shows the development of the centreline temperature along the channel axis computed for the three discussed cases. Because of non-corrected voltage the temperature in the "original" case is too high. The axial dependencies of the centreline temperatures obtained for the "corrected" and "modified" case are similar mutually, the slope of the latter being slightly lower.

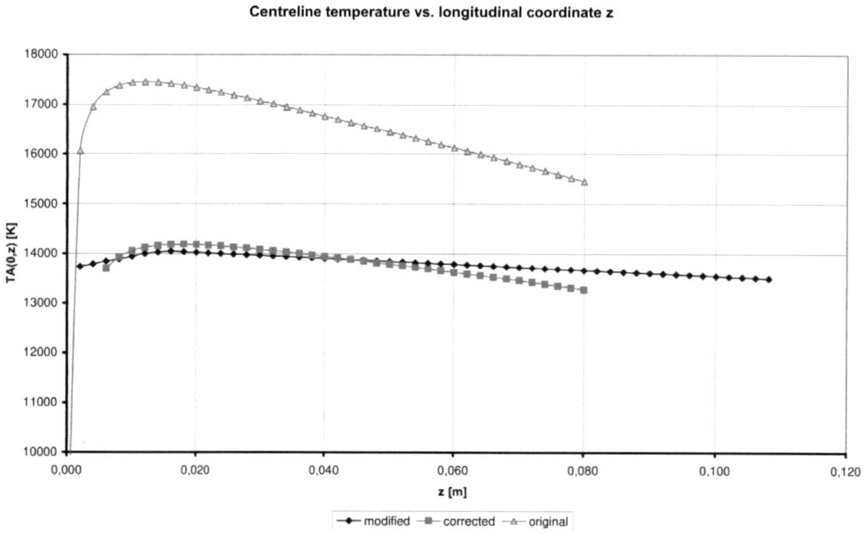

Figure 3. Comparison of the dependency of the centreline temperature $T_A(0,z)$ on the longitudinal coordinate z for the "modified", "corrected" and "original" case.

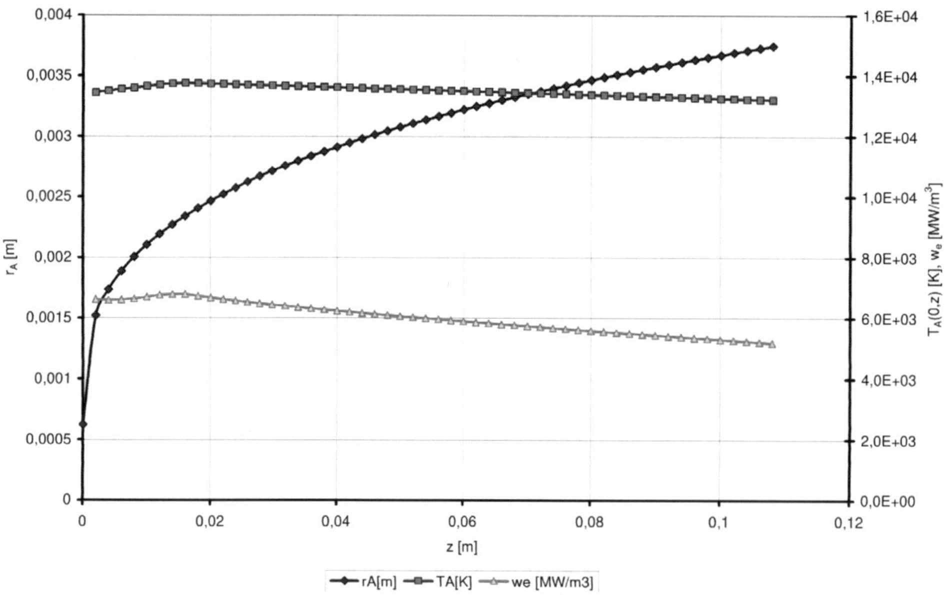

Figure 4. The dependencies of the arc radius r_A, the centreline temperature $T_A(0,z)$ and the arc specific radiated power w_e on the longitudinal coordinate for the "modified" case.

Figure 4 illustrates the connection among the specific radiated power $w_e[T_A(0,z)]=4\pi\varepsilon_N[T_A(0,z)]$, the corresponding temperature and the arc radius. The curve $w_e(z)$ is rather flat, with the small maximum near the channel input and the slight decrease towards the channel output. This tendency corresponds to the course of both other dependencies in figure 4: the net emission coefficient ε_N decreases with the decreasing temperature and with the increasing radius.

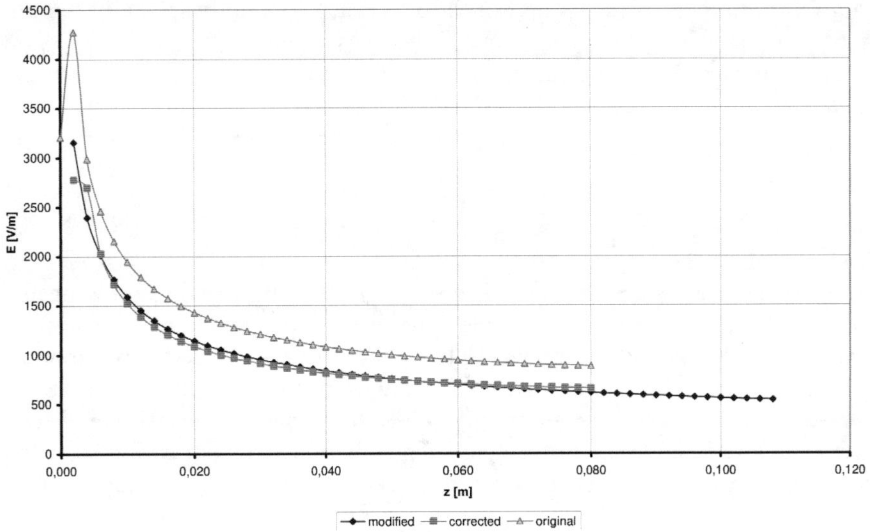

Figure 5. The dependency of the electric field strength E on the longitudinal coordinate z for the "modified", "corrected" and "original" case.

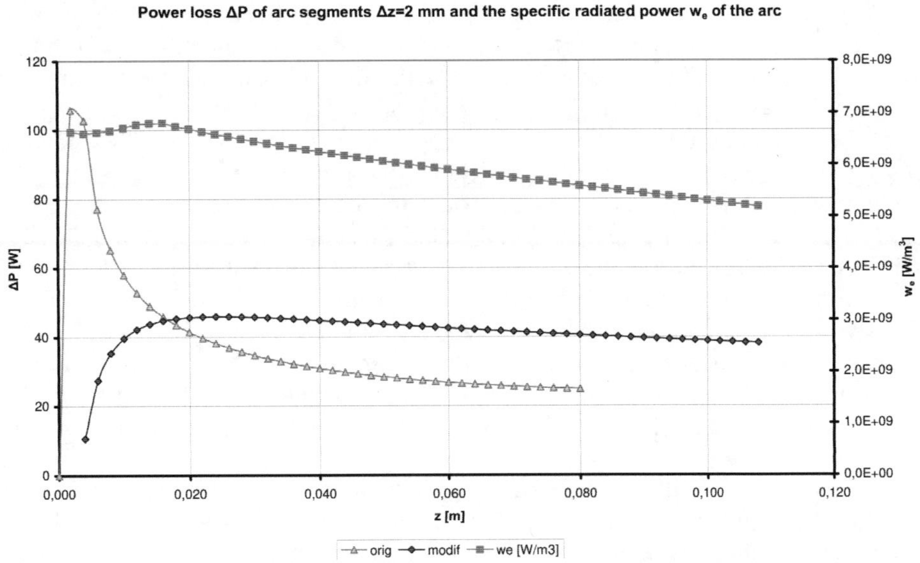

Figure 6. The axial development of the arc radiation losses ΔP of the cylindrical segments Δz=2 mm for the "original" and "modified" cases, and the specific radiated power w_e of the arc for the "modified" case.

The dependencies of the electric field strength E vs. longitudinal (axial) coordinate z in figure 5 are similar for all cases. In the "corrected" and "modified" case they are almost the same. The axial dependencies of the radiation loss ΔP of individual segments in the "original" and "corrected" case copy the course of $E(z)$. This fact is due to the computation used in these cases $\Delta P_r(z_i) = (1- \varepsilon)\, \Delta U(z_i)I$ as explained above.

In figure 6 the axial development of the arc radiation losses ΔP of the cylindrical segments $\Delta z = 2$ mm for the "original" and "modified" cases are given, together with the specific radiated power w_e of the arc for the "modified" case again. For the "original" case, the arc radiation losses ΔP exhibit the sharp maximum at the beginning of the channel, the rapid decrease in the middle part and almost constant value near the end. In the "modified" case, when the modified computation of the radiation loss using the net emission coefficient and the effective arc radius has been applied, the resulting dependency of the radiation losses of individual segments along the axis is different: excepting the area near the cathode, the radiation losses are distributed almost uniformly along the channel.

5. Conclusions

The paper presents the modified model of the electric arc burning in argon flowing through the arc heater's cylindrical channel and demonstrates the computed results obtained for real input data.

An older version of the model was updated. The comparison of the results computed from rough and carefully corrected new measured data clearly shows that the previous presumption of negligible power loss of cathode and anode spots is not valid. A proper segmentation of the channel and mathematical correction of input data is necessary. The obtained very flat corrected radial temperature profile indicates that the sometimes used two-zone models are probably not far from reality [1], [2], [3].

The minor modification concerns relations describing radial temperature dependencies. The terms of the input gas temperature T_0 have been inserted into approximations of $T_A(r,z)$ and $T_C(r,z)$ which is more correct but the actual influence on the results is weak. The substantial modification focuses on the radiation loss term in the energy conservation equation. The net emission coefficient of argon was used for the expression of the radiation loss. Because the real situation does not comply with the theoretical presumption of a homogeneous gas cylinder of constant temperature and infinite length, an effective radius $k_r r_A$ of the radiating gas cylinder was introduced as another computational parameter. Naturally, the designed modification is not exact but its advantage consists in preserving the connection of the computed radiation loss to the temperature and geometry of the radiating element. Due to this modification, the resulting dependency of the computed radiation loss of individual segments on the longitudinal coordinate z exhibits no significant extremes.

The computations using the measured integral quantities show the dominant role of radiation in the total power loss of the arc inside the arc heater's anode channel. The character of the obtained dependencies of the centreline temperature $T_A(0,z)$ and electric field strength $E(z)$ on the longitudinal coordinate z is very similar to the corresponding curves given in [2], [3], [4], [11].

From a practical point of view, also some modifications making the computation faster and easier have been prepared and found advantageous. Preliminary computations of integrals $F[T_A(0,z),n]$ (8) have been replaced by analytical computation with suitable approximations of temperature dependencies of argon properties and/or their products $\rho{\cdot}a{\cdot}h$, $\rho{\cdot}a$, σ. The total duration of the calculation can be reduced using pre-estimated computational parameters n_r, n based on previous experience. From the energy balance at the input and output cross-sections of the arc column approximate values of centreline temperatures can be simply estimated.

Acknowledgement

The research leading to these results has received funding from the European Community's Seventh Framework Programme (FP7/2007-2013) under grant agreement no. 230126. The research was also financially supported by the Ministry of Education of the Czech Republic under the projects no.

MSM0021630513 and MSM0021630516. The authors would like to express their gratitude to the anonymous reviewers for their time and effort spent during careful inspections as well as their helpful comments and suggestions for the improvement of the paper.

References

[1] Frost L S and Liebermann R W 1971 Composition and transport properties of SF_6 and their use in a simplified enthalpy flow arc model *Proc IEEE* **59** pp 474-85

[2] Hermann W, Kogelschatz U, Ragaller K and Schade E 1974 Investigation of a cylindrical, axially blown, high-pressure arc *J Phys* D *Appl Phys* **7** pp 607-19

[3] Lowke J J and Ludwig H C 1975 A simple model for high-current arcs stabilised by forced convection *J Appl Phys* pp 3352-60

[4] Lowke J J and Tuma D T 1975 Prediction of properties of arcs stabilised by forced convection *J Appl Phys* **46** pp 3361-67

[5] Ramakrishnan S, Stokes A D and Lowke J J 1978 An approximate model for high-current free-burning arcs *J Phys* D *Appl Phys* **11** pp 2267-80

[6] Aubrecht V and Bartlova M 2002 Net emission coefficients in argon arc plasmas *Czech J Phys* **52** D pp 522-7

[7] Aubrecht V Private communication 2009

[8] Gleizes A 2009 Transport coefficients and radiation in thermal plasma *Summer School of XVIIIth Symp on Physics of Switching Arc (Nove Mesto n Mor, Czech Rep)* ed V Aubrecht

[9] Heinz J and Senk J 2004 *Czech J Phys* **54** C pp 702-8

[10] Gregor J, Jakubova I and Senk J 2008 *Chem listy* **99** pp 1234-45

[11] Hermann W, Kogelschatz U, Niemeyer L, Ragaller K and Schade E 1974 Experimental and theoretical study of stationary high-current arc in supersonic nozzle flow *J Phys* D *Appl Phys* **7** pp 1703-22

Journal of Physics: Conference Series **275** (2011) 012008 doi:10.1088/1742-6596/275/1/012008

Plasma arc cutting technology: simulation and experiments

G Cantoro[1], V Colombo[1], A Concetti[1], E Ghedini[1], P Sanibondi[1], F Zinzani[1], F Rotundo[1], S Dallavalle[2] and M Vancini[2]
[1] Department of Mechanical Engineering (D.I.E.M.) and Research Center for Applied Mathematics (C.I.R.A.M.), Alma Mater Studiorum-Università di Bologna, Via Saragozza 8, 40123 Bologna, Italy.
[2] Cebora S.p.A., Via Andrea Costa 24, 40057 Cadriano di Granarolo, Italy.

E-mail: emanuele.ghedini@unibo.it

Abstract. Transferred arc plasma torches are widely used in industrial processes for cutting of metallic materials because of their ability to cut a wide range of metals with very high productivity. The process is characterized by a transferred electric arc established between an electrode inside the torch (the cathode) and another electrode, the metallic workpiece to be cut (the anode). In order to obtain a high quality cut and a high productivity, the plasma jet must be as collimated as possible and must have the higher achievable power density. Plasma modelling and numerical simulation can be very useful tools for the designing and optimizing these devices, but research is still in the making for finding a link between simulation of the plasma arc and a consistent prevision of cut quality. Numerical modelling of the behaviour of different types of transferred arc dual gas plasma torches can give an insight on the physical reasons for the industrial success of various design and process solutions that have appeared over the last years. Diagnostics based on high speed imaging and Schlieren photography can play an important role for investigating piercing, dross generation, pilot arcing and anode attachment location. Also, the behaviour of hafnium cathodes at high current levels at the beginning of their service life can been experimentally investigated, with the final aim of understanding the phenomena that take place during those initial piercing and cutting phases and optimizing the initial shape of the surface of the emissive insert exposed to plasma atmosphere.

1. Introduction

The PAC process [1] is characterized by a transferred electric arc that is established between an electrode, that is part of the cutting torch (the cathode) [2-6], and another electrode, that is the metallic workpiece to be cut (the anode). In order to obtain a high quality cut and a high productivity, the plasma jet must be as collimated as possible and must have the higher achievable power density. Plasma modeling and numerical simulation can be very useful tools for the designing and optimizing these devices [7-8, 11,13], but research is still in the making for finding a link between simulation of the plasma arc and a consistent prevision of *kerf* formation, to predict cut quality. In this frame, qualitative diagnostics based on high speed imaging and Schlieren photography can play an important role for investigating piercing, dross generation, pilot arcing and anode attachment location [9-10]. Also, the behaviour of hafnium cathodes at high current levels at the beginning of their service life can been experimentally investigated [14-15], with the final aim of understanding the phenomena that take

place during those initial piercing and cutting phases and optimizing the initial shape of the surface of the emissive insert exposed to plasma atmosphere.
State of the art knowledge in PAC is defined more by the huge amount of patents literature than by journal papers; this fact induces a strong need for understanding the physical reasons behind industrially patented successful ideas that, due to patenting rules and strategies, are often not completely and correctly described. The particular approach of this paper is in line with the research work done by the Authors in recent years to fulfil the abovementioned need, taking as a starting point either patented solutions or commercial solutions based on those patents in the field of plasma arc cutting.

2. Optimization of plasma arc cutting of mild steel thin plates

PAC processes of mild steel thin plates (thickness in the range 1-3mm), characterized by low current levels (25-45A) and the use of O_2 both as plasma gas and secondary gas, through a Cebora HQC Plasma Prof 164 plasma source, have been studied. The optimization of PAC of mild steel thin plates, both in terms of cut quality and performances of the consumables, have been accomplished in order to achieve cut quality standards and productivity levels usually obtainable through laser cutting processes. The optimization process has been carried out through the simultaneous planning and analysis of experimental tests and numerical simulations, by means of a 2-d FLUENT-based numerical model [7-8].
All the accomplished experimental tests have focused on the investigation of phenomena tip-ically considered to be critical for the cutting process in those conditions: unevenness of the kerf surface, inclination of the cut surfaces and dross formation. The influence of operative parameters on critical phenomena has been studied, mainly changing secondary gas flow and the plasma gas inlet pressure. Also, a better design of consumables, in particular nozzle, electrode and diffusers, and the optimization of current profiles, in particular pilot arc current levels, have been carried out. Modifications that have been finally introduced to the operative parameters and to the torch configuration are:

- the reduction of the secondary gas flow rate (lower than 9 lpm);
- a considerable improvement of the plasma flow rate (from 4 lpm to 6.5 lpm);
- the introduction of a modified nozzle with the conical shape of the orifice end, characterized by an optimum value of the depth of the conical area;
- the introduction of an optimized diameter of the hafnium insert (1.07 mm);
- a reduction of the diameter of the holes of the plasma gas diffuser (0.4 mm);
- the optimization of the current profile during pilot arcing.

Modelling and numerical simulation allowed a better understanding of the physical phenomena concerning the critical aspects initially pointed out and enabled us to detect successful design solutions. In particular, simulation results showed that all modifications that have been introduced, both in terms of operating parameters and consumables geometry, lead to:

- an increase of the axial plasma temperature inside the nozzle (Fig. 1);
- an increase of the plasma swirl velocity in the nozzle orifice region (Fig. 2);

the removal of turbulent vortexes in the nozzle orifice exit region (Fig. 3).
The integration of the results of these two activities allowed to overcome the critical aspects initially pointed out, improving plasma jet constriction and nozzle cooling and reducing plasma jet instabilities, leading to the improvement of cut quality and the reduction of erosion phenomena of the torch consumables, in particular of the nozzle.

3. High Speed Imaging of transient phenomena and Schlieren photography in PAC

High speed camera (HSC) imaging of pilot arcing and piercing process phases with dual gas torches have been investigated using mild steel and stainless steel plates of various thickness, in various operating conditions. Images have been obtained using a NAC Memrecam K3R HS camera with a 180 mm focal-length lens, protected by a sacrificial neutral filter at a stand-off distance of 1 m from a Cebora (HQC164 for the range 25-120 A and HQC254 for the range 120-250 A) plasma torch, joined

Journal of Physics: Conference Series **275** (2011) 012008 doi:10.1088/1742-6596/275/1/012008

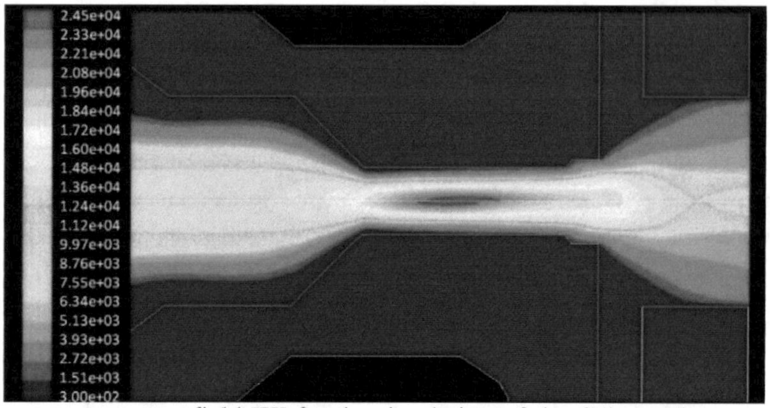

Figure 1. Plasma temperature field [K] for the simulation of the Cebora plasma torch under the following operating conditions: 40 A; plasma gas pressure (O2): 540 kPa (upper half) and 590 kPa (lower half); secondary gas pressure (O2): 40 kPa; nozzle diameter: 0.8 mm.

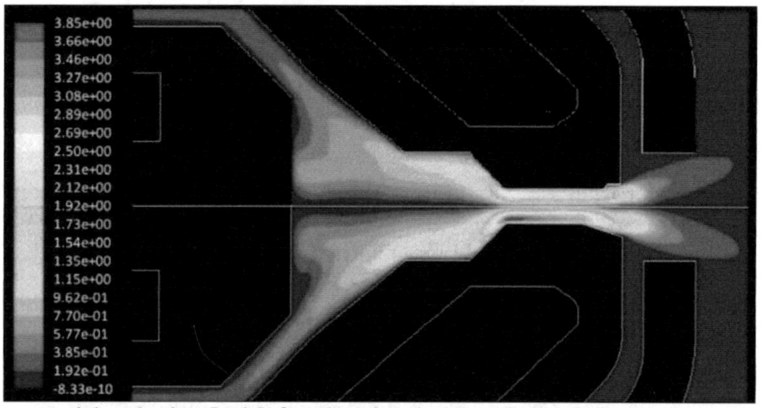

Figure 2. Plasma swirl velocity [m/s] for the simulation of the Cebora plasma torch under the following operating conditions: 40A; plasma gas pressure (O2): 540 kPa; secondary gas pressure (O2): 40 kPa; nozzle diameter: 0.8mm with a nozzle orifice end cylindrically shaped (upper half) and conically shaped (lower half).

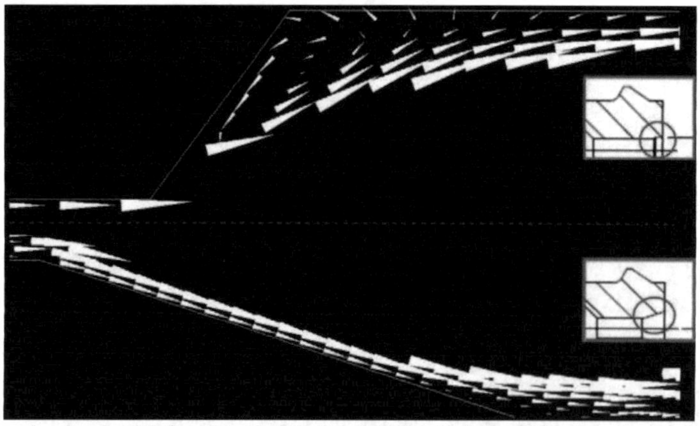

Figure 3. Plasma velocity [m/s] for the simulation of the Cebora plasma torch under the following operating conditions: 40 A; plasma gas pressure (O2): 540 kPa; secondary gas pressure (O2): 40 kPa; nozzle diameter: 0.8 mm with the nozzle orifice end cylindrically shaped (upper half) and conically shaped (lower half).

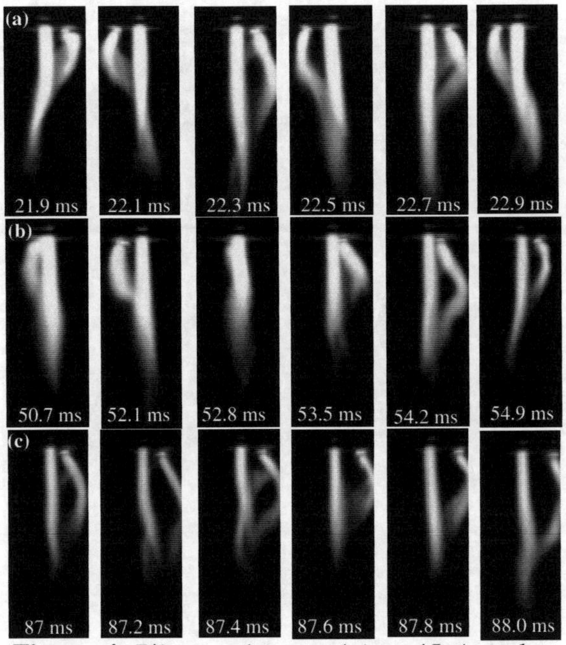

Figure 4. Pilot arc images *(a)* at 45 A and at constant time steps of 0.2 ms, *(b)* during current slope from 45 A to 25 A and at constant time steps of 0.7 ms, *(c)* at 25 A and at constant time steps of 0.2 ms.

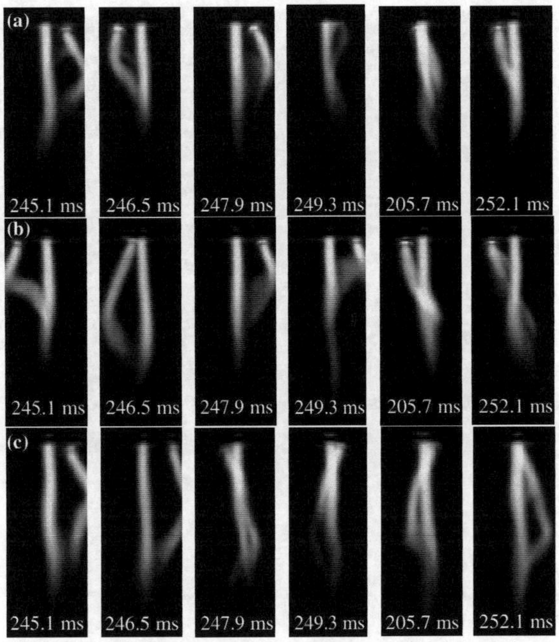

Figure 5. Pilot arc images at 25 A and at constant time steps of 1.4 ms with plasma gas flow rate of about: *(a)* 21 slpm, *(b)* 28 slpm, *(c)* 36 slpm.

with a digital oscilloscope (LeCroy LT374M). Synchronizing the start acquisition time of both instruments, the images captured by the HSC have been time-correlated to the oscilloscope waveforms of voltage, current and pressure inside the nozzle. Pilot arcing images have been obtained with air as pilot arc plasma gas, with a 1,9 nozzle diameter and in the absence of the torch shield cup (without secondary gas) in order to observe the arc loop attachment on the nozzle tip. Pilot arcing duration has been extended with respect to real cutting conditions in order to fully observe arc loop attachment phenomena. All images haven been acquired at 10000 fps and 1/200000 s shutter time without any filtering.
The technique has provided new insight of the PAC process and some interesting phenomena have been highlighted.

3.1. Effects of different operating parameters on pilot arc behaviour

A study of the influence of different parameters on the rotation of the arc loop attachment on the nozzle tip during pilot arcing has been carried out through a series of experiments at different operating conditions. The influence of the current level on pilot arcing has been studied setting a current waveform with two different current levels: a first phase at 45A, with a second phase at 25A. As shown in Fig.4, an increase in arc current level induces an increase of both the pilot arc cross section and its visible light emission, while inducing an enhanced random jumping arc attachment behavior on the nozzle tip. Fig. 5 shows three high speed imaging experiments carried out with different plasma gas flow rates (21, 28 and 36 slpm). This comparative analysis has shown that an increase in the flow rate leads to the reduction in the pilot arc rotational speed on the nozzle tip; the rotational speeds have been evaluated analyzing the single frames for the three cases and the frequencies are about 2000 Hz, 1000 Hz and 400 Hz, respectively. A series of results that will not be detailed here shows how different swirl strength conditions for the plasma gas flow also have an impact on the rotation of the arc root attachment. In the case of higher swirl strength the arc loop

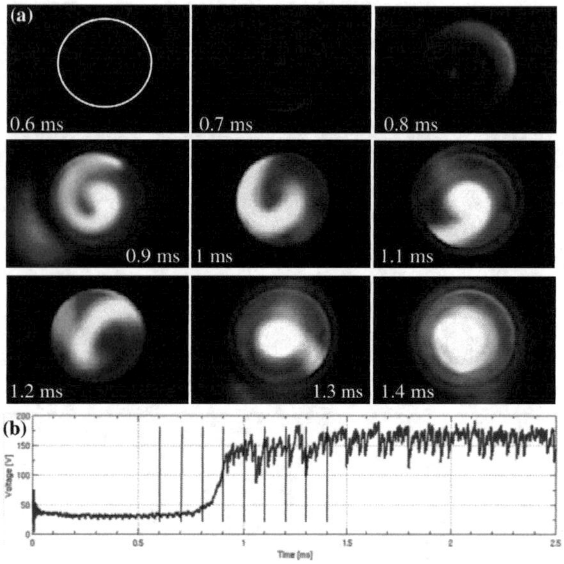

Figure 6. *(a)* Sequence of pilot arc images during an arc current transient subsequent to arc ignition. *(b)* Images are time-correlated with the oscilloscope waveform of the voltage between the electrode and the nozzle. In the first image frame the position of the shield hole is shown.

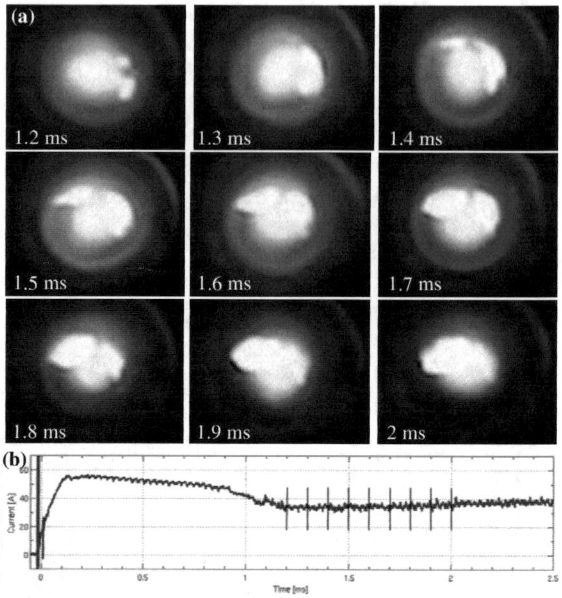

Figure 7. *(a)* Sequence of pilot arc images during an arc current transient subsequent to arc going out of the nozzle. *(b)* Images are time-correlated with the oscilloscope waveform of arc current.

attachment experiences an entire (360°) rotation along the nozzle tip without any erratic or discontinuous moving. The total time for a complete rotation of the arc around the nozzle tip is 11.5 ms. The case of lower swirl strength is characterized by erratic and discontinuous movements of the arc attachment on the nozzle tip experiencing an incomplete (180°) rotation on the nozzle tip.

3.2. Cathode surface images during pilot arcing

Visualization of the cathode surface during pilot arcing with high time resolution was accomplished by positioning the torch and the camera lens on the same horizontal axis. Four different sequences (with different focal ratios in the camera set-up) of pilot arc images coming from four different experiments carried out in the same operative conditions have been obtained. Realistic conditions for the cutting of MS plates with 250 A and O_2/air as the plasma/secondary gas have been used. Every sequence of images shows at best a particular phenomenon that takes places during a pilot arcing.

Images of Fig. 6 were obtained with a focal ratio of f/16; they concern the phase in which the pilot arc, once initiated, is blown out of the nozzle by the plasma gas flow, with a swirling motion due to the swirl component of the plasma flow. The increase in the voltage can be associated to the increase in the arc axial length.

Images of Fig. 7 were obtained with focal ratio of f/32 and they refer to the phase in which the arc has already exited the nozzle, creating an arc loop protruding out of it. The bright areas between the arc root and the copper holder of the electrode, that appear in some frames, are interpreted as electron emission spots on the molten hafnium surface of the electrode; their movement is ruled by the drag of the swirling plasma gas flow.

Images of Fig. 8 were obtained with focal ratio of f/5.6, aimed at obtaining information on conditions on the nozzle tip. The torch operative conditions used in these movies are the same as those

Journal of Physics: Conference Series **275** (2011) 012008 doi:10.1088/1742-6596/275/1/012008

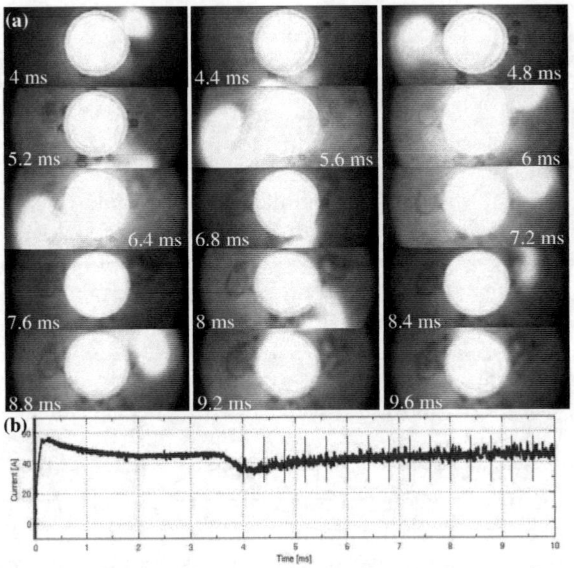

Figure 8. *(a)* Pilot arc images at constant time intervals and at about 45 A. *(b)* Images are time-correlated with the oscilloscope waveform of arc current.

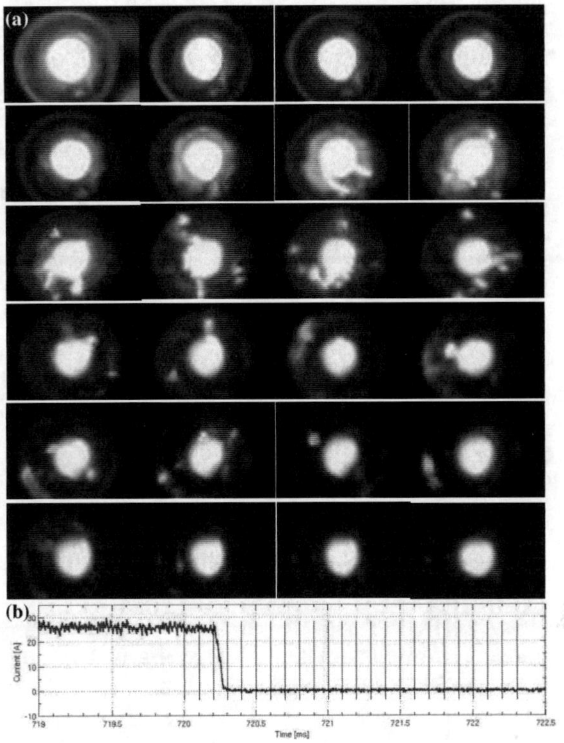

Figure 9. *(a)* Sequence of pilot arc images, from 720 ms to 722,3 ms, at constant time steps of 0,1 ms, during arc current shutdown from 25 A to 0 A. *(b)* Images are time-correlated to the arc current oscilloscope waveform.

characterized by lower swirl strength and induce an intermittent motion of the arc loop attachment on the nozzle tip. The intermittent root attachment of the pilot arc induces the formation of dark dots on the nozzle tip, even though the arc maintains its non-moving and localized attachment on the same point for less than 0.4 ms.

Images of Fig. 9 were obtained with focal ratio of f/5.6 and with a 2,5 mm diameter nozzle while in Fig.s 6, 7 and 8 the nozzle diameter was 1.9 mm. Images obtained during this experiment show both the detachment from the hafnium molten pool of molten hafnium particles and their trajectories. The HS images clearly show that the presence of centrifugal forces due to the plasma gas swirling, plays a leading role in electrode erosion. The time correlation between HS images and current waveform shows that the detachment of molten hafnium particles from the emissive surface begins when the arc current drops to zero with current going from 25 A to 0 A in less than 0.1 ms.

3.3. Images of double arcing in transferring arcs

Fig. 10 shows images of a realistic phase of arc transferring during the piercing of a MS plate 20 mm thick at 200 A and O2/Air as plasma/secondary gas, respectively. Gas flow rates and torch stand-off have been set to increase the probability of double arcing taking place during piercing. The sequence of images shows different phenomena that can be associated to the double arcing. In particular green vapours and silver grey vapours appear in most of the image frames and can be correlated to copper (due to arc attachment on the nozzle

and on the shield) and hafnium (due to arc root attachment instability on the electrode emissive surface) vapour emissions, respectively.

The hypothesis of arc root attachment instability is also supported by the evidence in some of the frames of Fig. 10 of traces of solid particles (probably hafnium oxide) exiting the nozzle. Some other image frames show the presence of an arc attachment on the nozzle, the arc having disappeared from its usual position along the torch axis. The correspondence between arc instabilities showed by the images and the spikes in the oscilloscope waveform of arc voltage leads to a hypothesis where the arc voltage between electrode and nozzle can be assumed as an arc stability indicator: the isolated peaks can be related to hafnium vapors emission while continuous spikes to arc attachment on the nozzle and the shield, respectively.

Non-destructive double arcing phenomena during piercing probably occur as a consequence of the deposition of a small amount of hafnium oxide on the nozzle orifice wall, inducing a local increase of the radial electric field and an increase in the probability of double arcing. When double arcing occurs, rapid [<1 ms] evaporation of the hafnium oxide with vapor emission from the nozzle restores the previous and normal arc behavior. Due to their very short duration, the described phenomena can assume a non-destructive character.

Two more phenomena, not detailed here, were investigated by some of the authors [10]. The first study concerning the tracking of trajectories and velocities of ejected particles from the nozzle orifice during pilot arcing, on a shorter time-scale than the one typical for the ejection of liquid metal droplets during transients [4], probably coming from cracking of the solid layer of HfO_2 on the emitter during arc re-start [1].

The second one concerning the influence of the level of alignment of torch head components on the plasma arc behavior during piercing phases. In particular, it was found that a relative non-alignment between shield and electrode grater then 0.025 mm can induce destructive piercings at high current (>200 A). Schlieren photography has also been used by some of the authors to visualize density gradients due to turbulence and temperature in real cutting conditions, in order to investigate plasma and secondary gas interactions and their impact on cut quality with feedback on torch design

4. Experimental analysis of the behaviour of high current electrodes in pac during first cycles

Cathode erosion phenomena put still a limit in the improvement of the performances of PAC process, with respect to its main competitors; in particular for the operative conditions in which oxidizing plasma gases (oxygen and air) and high current levels (> 200 A) are used in presence of an Hafnium (Hf) emitter. Several studies [2-6] have been accomplished in order to understand phenomena that give rise to Hf erosion in different phases of the PAC operative cycle.

The behaviour of Hf cathodes at the beginning of their service life when operating at high current levels in the PAC process has been experimentally investigated with the final aim of understanding the phenomena that take place during those initial piercing and cutting phases and optimizing, with respect to expected service life, the initial shape of the surface of the emissive insert exposed to plasma atmosphere. Hf erosion mechanisms on the emitter surface, modifications of the Hf insert morphological structure, together with the effect of deposition of molten emissive material on the nozzle inner surface have been studied in realistic operative conditions for cutting mild steel plates with oxygen as plasma gas and air as shield gas at relatively high current level (250 A).

A new experimental procedure have been proposed, starting the investigation with an initially plane insert emission surface, subsequent cutting phases have been accomplished checking each time the dimension of the growing concave recess on the insert and the amount of Hf oxide collected on the nozzle surface during the first few cutting cycles while using these information for defining subsequent optimization steps.

Morphological and compositional analysis of the tested electrodes and nozzles led to a detailed description of Hf erosion phenomena and of the trend of modification of the emission surface and of the interface between the Hf insert and the copper holder.

High speed imaging during piercing phases has been used to highlight the possible presence of non destructive double arcing phenomena taking place at the beginning of electrode-nozzle service life as a consequence of deposition of Hf based material on the nozzle. Also for highlighting the different behavior during start-up phases of new or used electrodes.

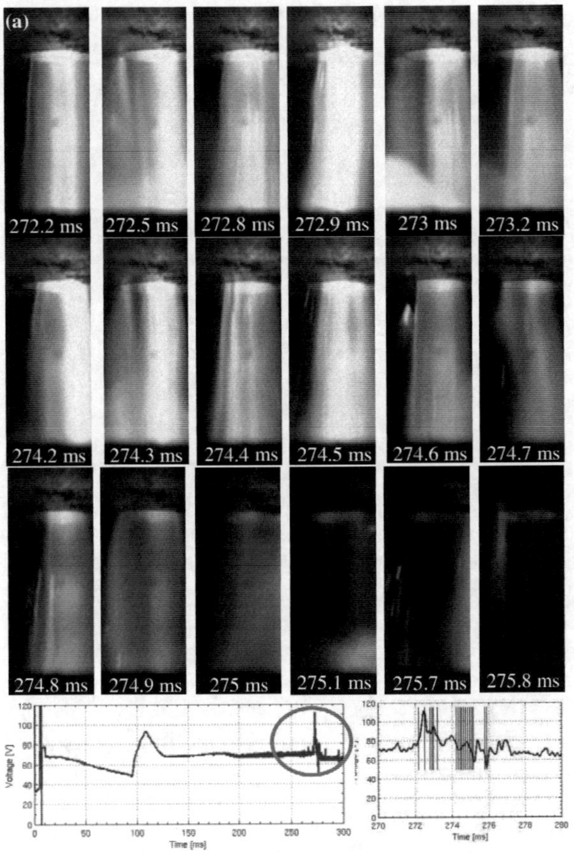

Fig. 10 *(a)* Transferring arc images, at different time steps, for a torch with 1.6 mm nozzle diameter and 3 mm shield diameter, at 200A. *(b)* Images are time-correlated to the oscilloscope waveform of the electrode-nozzle voltage.

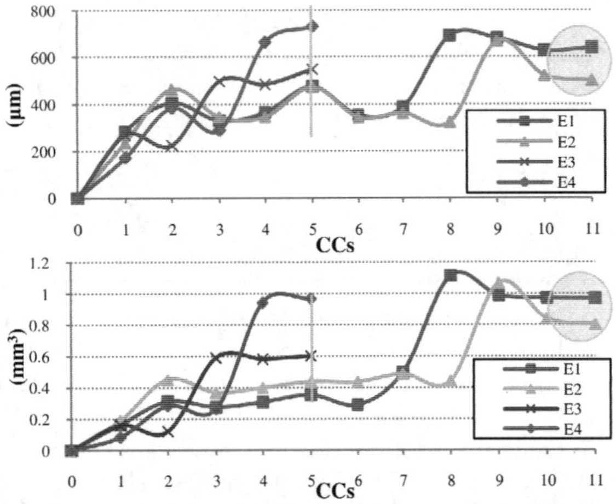

Fig. 11 Trends *(a)* of the maximum recess depth and *(b)* of the calculated eroded volumes, as a function of the number of CCs, for electrodes with no initial recess

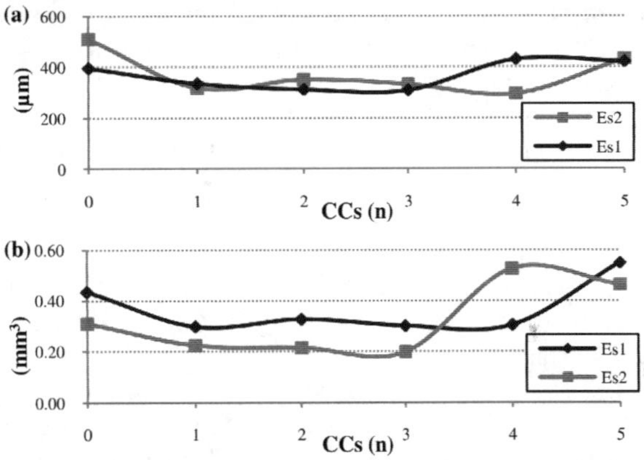

Fig. 12 Trends *(a)* of the maximum recess depth and *(b)* of the measured eroded volumes, as a function of the number of *CCs*, for electrodes *Es1* and *Es2*

4.1. Experimental procedure

In general terms, the optimal shape of the initial recess of the Hf insert is strongly dependent upon "typical operating and geometric conditions" of the torch being optimized. For these reasons, an experimental procedure is proposed here, starting the investigation with an initially plane insert emission surface and analyzing the evolution of the recess depth naturally created during the first few cutting cycles (*CCs*) while using these information for defining subsequent optimization steps.

The proposed experimental procedure can be divided into different stages. The first stage is characterized by a series of tests accomplished on electrodes having an initially plane insert emission surface. The erosion tests were stopped after every cutting cycle to check the Hf eroded volume and the overall nozzle condition. Modifications of the Hf insert morphological structure were also monitored and this first test was stopped when the emissive surface morphology reached a stabilization phase as a function of the number of *CCs*. The dimension of the recess that was naturally created on the emissive surface at the end of the first stage was taken as reference point to design some intermediate spherical recesses to be tested on erosion during *CCs* in the second stage of the procedure. The Hf eroded volume, the overall nozzle condition and the Hf insert morphological structure evolution over *CCs* were monitored, in order to identify the stabilization phase for each tested electrode. The modifications occurred on the emissive surface morphology of tested electrodes allowed to identify the optimal spherical recess.

4.2. Behaviour and analysis of electrodes with no initial recess

Four experimental tests (*E1÷E4*) were accomplished. For each test, 1 electrode with no initial recess underwent 5 *CCs*; a new nozzle was used for every *CC*. 3-D topographies and volume measurements were accomplished for each electrode after every *CC*. After the first arc ignition, all the tested electrodes evidenced the presence of a concave recess in the Hf emissive insert and an annular zone with Hf deposit surrounding the insert boundaries on the cathode surface. Moreover, SEM macrographs exhibit the presence of a fine dispersion of Hf oxide surrounding the massive Hf deposit, together with HfO_2 droplets. Finally, part of the Hf based particles ejected from the insert is deposited on the inner nozzle surface.

The evolution of the eroded volumes (Fig. 11 (a)) at the first stage (which show a trend similar to the one of the maximum cavity depth, Fig. 11 (b)) suggests how the erosion behaviour of each insert is only partially determined by the morphology of the emitting surface as measured at the previous step. Electrodes E1 and E2, between 2 and 5 *CCs*, maintained their cavity depth and erosion volume almost constant, while no or extremely limited Hf emission and deposition on the nozzle inner surface was detected proving how the Hf volumes eroded during first 2 *CCs* could be removed from an optimized emissive insert. Due to this particular behaviour, they have been tested for an increased number of starts, in order to better study the evolution of the recess spontaneously established, and they have been considered as a reference point to start an experimental procedure for the optimization of the recess shape.

Electrodes E1 and E2, subjected to 11 *CCs*, are both characterized by massive Hf ejection events after the first 5 *CCs*, in particular: the eroded volumes skip from 0.31 – 0.43 mm^3 to 1 mm^3 and the maximum recess depth reaches a level of about of 0.7 mm (Fig.11). The levels of eroded volume and maximum recess depth that electrodes E1 and E2 reach, in a stabilized manner, after 11 *CCs* are similar to those ones that electrodes E3 and E4 reached, with an non linear trend, after 5 *CCs*. They both reach the stabilization of the emissive surface morphological structure after 9 *CCs*.

4.3. Behaviour of electrodes with initially optimized recess

On the basis of the results obtained from tests on electrodes E1 and E2 after 5 *CCs*, two (non-trial) spherical recess shapes have been designed:

- from electrode E1 a spherical recess with an initial recess depth of about 400 μm and an initial recess volume of 0,31 mm^3 (electrode Es1);
- from electrode E2 a spherical recess with an initial recess depth of about 500 μm and an initial recess volume of 0,43 mm^3 (electrode Es2).

Experimental evidences showed that both for electrode Es1 and Es2 the eroded volume levels and the recess depth were significantly reduced with respect to the ones reached on E1 and E2 after 10 *CCs* (Fig. 12). Moreover, both eroded volume and maximum recess depth levels of Es1 and Es2 reach a similar value after accomplished erosion tests, although fresh electrodes were designed with a slightly different initial shape of the spherical recess. This result enabled us to consider the recess shape obtained at this stage with electrodes Es1 and Es2 as a reference point to design the optimized spherical recess.

Experimental results showed that an initial planar emission surface gives rise to massive Hf ejections during first cutting cycles, inducing a preferred concave shape on the emitter, with consequent massive Hf deposition on the inner nozzle surfaces that negatively affect cut quality and nozzle service life. Moreover, the developed experimental procedure showed that the optimization of the initial recess shape of the Hf emitter surface not only minimizes the deposition of Hf oxides on the nozzle inner surfaces maintaining an electrode service life comparable to the one of electrodes with initially planar emission surface, as affirmed in [6], but also positively affects the subsequent trend of the Hf erosion rate, improving electrodes service life on the whole.

5. References

[1] V A Nemchinsky and W S Severance "What we know and what we know not about plasma arc cutting", J. Phys. D: Appl. Phys., **39,** R423-R438, 2006.

[2] V A Nemchinsky and M S Showalter "Cathode erosion in high-current high-pressure arc", J. Phys. D: Appl. Phys., **36** 704-712, 2003.

[3] V A Nemchinsky "Cyclic erosion of a cathode in high-pressure arcs", J. Phys. D: Appl. Phys., **36,** 1573-1576, 2003.

[4] J Peters, F Yin, C F M Borges, J Heberlein and C Hackett "Erosion mechanism of hafnium cathodes at high current", J. Phys. D: Appl. Phys., **38,** 1781-1794, 2005.

[5] L Luo and R W Jr Couch "Electrode for plasma arc torch", Hypertherm Inc. Patent US 5,464,962, (7 November 1995).

[6] L Luo and R W Jr Couch "Electrode for plasma arc torch", Hypertherm Inc. Patent US 5,601,734, (11 February 1997).

[7] V Colombo, E Ghedini, G Masini, D Russo, S Dallavalle and M Vancini "2-D And 3-D Fluidynamic And Plasma Characterization Of Dc Transferred Arc Plasma Torches For Metal Cutting", High Technology Plasma Processes 9 - HTPP9, St. Petersburg, 2006 and Journal of High Temperature Material Processes, **10**, pp. 379 – 392, 2006.

[8] V Colombo, A Concetti, E Ghedini, S Dallavalle and M Vancini, "Understanding Plasma Fluid Dynamics Inside Plasma Torches Through Advanced Modeling", IEEE Trans. Plasma Sci., **36**, 389-402, 2008.

[9] V Colombo, A Concetti, E Ghedini, S Dallavalle and M Vancini, "High Speed Imaging of Pilot Arcing and Piercing in PAC", IEEE Trans. Plasma Sci, **36**, 4 part 1, 1042-1043, 5th Triennial Special Issue on Images in Plasma Science, 2008.

[10] V Colombo, A Concetti, E Ghedini, S Dallavalle and M Vancini "High speed imaging in plasma arc cutting: a review and new developments" (review paper), Plasma Sources Science and Technology, **18**, 1 - 24, 2009.

[11] V Colombo, E Ghedini, A B Murphy, M I Boulos, P Fauchais, A Gleizes, J Mostaghimi, P Proulx, D C Schram, "Adavanced thermal plasma modeling", High Temperature Materials Processes (review paper), **12**, 255-336, 2008.

[12] V Colombo, A Concetti, S Dallavalle, R Fazzioli, E Ghedini and M Vancini "Optimization of plasma arc cutting of mild steel thin plates", Journal of High Temperature Material Processes, **13**, issue 3-4, 267-285, 2009.

[13] V Colombo, P Sanibondi, E Ghedini "Thermodynamic and Transport Properties of H35 and F5 Plasma Cutting Mixtures in Non-Equilibrium", Journal of Physics D: Applied Physics, **42**, 1-24, 2009.

[14] V Colombo, A Concetti, E Ghedini, S Dallavalle, M Vancini, F Rotundo, C Chiavari "Experimental analysis of the behaviour of high current electrodes in plasma arc cutting during first cycles", 19th International Symposium on Plasma Chemistry - ISPC19, Bochum, 2009.

[15] V Colombo, A Concetti, E Ghedini, S Dallavalle and F Rotundo "Experimental analysis of the behaviour of high current electrodes in plasma arc cutting during first cycles", submitted to Plasma Sources Science and Technology.

Acknowledgments

Funding from D.I.E.M./Cebora S.p.A. research contract on project MIUR-D.M.28562 (Art.12EMec) and UniBo Strategic Project 2006-PLASMACUT is acknowledged.

Journal of Physics: Conference Series **275** (2011) 012009 doi:10.1088/1742-6596/275/1/012009

Mean absorption coefficients of air plasmas

N Bogatyreva, M Bartlova and V Aubrecht

Brno University of Technology, Faculty of Electrical Engineering and Communications, Technicka 8, 616 00 Brno, Czech Republic

E-mail: bartlova@feec.vutbr.cz

Abstract. Mean absorption coefficients of air (both Planck and Rosseland mean) have been computed for different splitting procedures of the whole frequency interval (10^{13} – $6\text{x}10^{15}$ s^{-1}) and various temperatures (1 000 – 30 000 K). Results have been compared and discussed to find the optimal values of mean absorption coefficients for the multigroup method of solution of the equation of radiation transfer.

1. Introduction

The propagation of radiation through plasma is an integral part of the description of the plasma. In the description of radiation transfer, mathematical modeling is of great importance. However, the non-linearity of the equations describing the radiation field and the strong dependence of the parameters and coefficients on the radiation frequency makes mathematical plasma models very complicated. Therefore, various approximate methods are used (method of net emission coefficient [1-3], method of partial characteristics [4, 5]).

One of the procedures for handling the frequency dependence of the coefficients in the equation of transfer is the multigroup method, which assigns a given photon to one of G frequency groups. All photons within a given group are treated in the same way, assigning average properties, such as the absorption coefficient, to that group. In practice, the average absorption coefficient is generally taken as either a group Rosseland or group Planck mean (depending on the absorption properties of the medium for the given frequency group).

In this paper, attention has been given to the absorption properties of air at the pressure of 1 atm and in the temperature range (1 000 - 30 000 K). Both Rosseland and Planck averages have been computed for different splitting procedures of the whole frequency interval (10^{13} - $6\text{x}10^{15}$ s^{-1}).

2. Radiation transfer

The main quantity describing the radiation transfer is the radiation intensity I. It is defined as the radiation power per unit solid angle and unit area of a surface that is placed normal to the rays. We distinguish between spectral and total intensity, I_ν and I, respectively:

$$I(\vec{r},\vec{\Omega})=\int_0^\infty I_\nu(\vec{r},\vec{\Omega},\nu)\,d\nu \quad . \tag{1}$$

Here, $\vec{r}$ is a position vector fixing the location of a point in space, and $\vec{\Omega}$ a unit direction vector.

In case of thermodynamic equilibrium, the radiation intensity is direction-independent, and given by the Planck function

$$B_\nu(\nu,T)=\frac{2h\nu^3}{c^2\left[\exp\left(\frac{h\nu}{kT}\right)-1\right]}. \tag{2}$$

A light beam traveling through a gas layer of thickness ds loses intensity by absorption and by scattering of photons. The attenuation of radiation is proportional to the magnitude of the incident energy (intensity I_ν) and to the length of the path

$$dI_\nu=-\kappa_\nu I_\nu ds\ . \tag{3}$$

Because scattering of photons by molecules and atoms is always negligible for heat transfer, the proportionality constant κ_ν represents the absorption coefficient.

In the local thermodynamic equilibrium the following relation holds between the emission and absorption coefficients

$$\varepsilon_\nu=\kappa_\nu B_\nu\ . \tag{4}$$

The complete stationary equation of radiation transfer for an absorbing and emitting medium is

$$\frac{dI_\nu}{ds}\equiv\vec{\Omega}\cdot\mathrm{grad}\,I_\nu=\kappa_\nu(B_\nu-I_\nu). \tag{5}$$

3. Absorption properties of plasmas

Radiation in arc plasmas depends, besides others physical quantities, on the concentrations of all chemical species in the plasma. In the calculation of the plasma a dry air composition was assumed at the pressure of 1 atmosphere consisting of N_2, O_2, Ar, and CO_2. The equilibrium composition of the air plasma was computed using Tmdgas code [6]. Atoms and up to the triple ions of N, O, Ar, C, and diatomic molecules O_2, N_2, N_2^+, NO, NO^+, CO, CO^+, and CN were assumed to be present.

When a photon interacts with a gas molecule, atom, or ion, it may be absorbed raising the particle's energy level. Conversely, a gas particle may spontaneously lower its energy level by the emission of an appropriate photon. There are three different types of radiative transitions that lead to a change in the energy level of a gas particle by photon emission or absorption:

- transition between non-dissociated („bound“) atomic or molecular states, called *bound-bound transitions* (bb)
- transitions from a „bound“ state to a „free“ (dissociated) one (absorption) or from „free“ to „bound“ (emission), called *bound-free transitions* (bf)
- transitions between two different „free“ states, *free-free transitions* (ff)

The total absorption coefficient is given as the linear sum of all three processes mentioned above

$$\kappa(\nu,T,p)=\kappa^{ff}(\nu,T,p)+\kappa^{bf}(\nu,T,p)+\kappa^{bb}(\nu,T,p)\ . \tag{6}$$

The calculation of the absorption coefficient represents a formidable task when experimental data is lacking, since the radial wave functions of all free and bound electronic states must be known. However, simplifications can be made by using various semi-empirical methods and hydrogenic approximations

3.1. Continuous radiation

The radiation continuum may be divided into photon emission from free-bound (recombination) radiation and free-free (bremsstrahlung) radiation.

Journal of Physics: Conference Series **275** (2011) 012009 doi:10.1088/1742-6596/275/1/012009

Since the kinetic energy of free electron is not quantized, its recombination with positive ion of a chemical species „*a*" leads to continuous radiation. Thus, for recombination radiation

$$h\nu = \frac{1}{2}m_e v_e^2 + E_\infty^a - E_i^a \ . \tag{7}$$

Here ν is the frequency of recombination radiation, h Planck's constant, v_e the velocity of free electron, E_∞^a the ionization potential of the atom or ion, E_i^a the energy of the *i*-th electronic energy state, where the electron is captured.

The spectral absorption coefficient of process (7) is related to the photon absorption cross section $\sigma_{\nu,i}^a$ by

$$\kappa_{\nu,i}^{bf} = \sigma_{\nu,i}^a . N_i^a \tag{8}$$

where N_i^a is the population density of the *i*-th electronic state of the absorbing species „*a*".

Neutral atoms and up to the triple ions of N, O, Ar, C elements were considered. Photo-ionization cross sections for the ground states of neutral atoms and ions of N, O, Ar, and C were calculated using analytic fits of theoretical cross sections from the "Opacity Project" [7]. The cross sections for excited levels of neutral atoms were calculated using the quantum defect method of Burges and Seaton [8]; this was applied to oxygen – 3s, 3p, 4s, and 4p states, nitrogen – 3s and 3p states, carbon – 3s, 3p, 4s, and 4p states. All other excited states of N, O, Ar, C were treated using the Coulomb approximation for hydrogen-like species [9]. The energy levels tabulated in [10] were used for all atoms and ions under consideration. We have taken into consideration almost 600 energy levels.

The hydrogen-like approximation was used for the treatment of free-free transitions, too. In contrast to bound electrons of a given level, free electrons contribute to the absorption coefficient at all frequencies. The relative importance of bound-free and free-free absorption depends on the state of ionization. At low temperatures, there are very few free electrons, and bound-free absorption dominates. At very high temperatures, most electrons are free, and free-free absorption is dominant.

3.2. Discrete radiation

Quantum mechanics postulates that the energy levels for atomic or molecular electron orbit as well as the energy levels for molecular rotation and vibration are quantized; i.e. , electron orbits and rotational and vibrational frequencies can only change by certain discrete amounts. Therefore, in bound-bound transitions, photons must have a certain frequency (or wavelength) in order to be captured or released, resulting in discrete spectral lines for absorption and emission

$$h\nu_{mn} = \frac{hc}{\lambda_{mn}} = E_m - E_n \ . \tag{9}$$

for the transition from the state *m* to the state *n*.

However, no spectral line can be truly monochromatic; rather, absorption or emission occurs over a tiny but finite range of wavelengths. The results are broadened spectral lines that have their maxima at the wavelength predicted by equation (9).

Numerous phenomena cause broadening of spectral lines. The most important are natural line broadening, Doppler broadening, Stark broadening, resonance broadening, and Van der Waals broadening .

Absorption coefficient κ^{bb} for discrete spectrum must take account of all overlapping spectral lines of all atoms and ions of the plasma. The absorption coefficient of an isolated line "*i*" of species "*a*" is [9]

$$\kappa_{a,i}^{bb}(\nu,T,p) = \pi r_0 c f_{mn}^{a,i} N_n^a(T,p) P_\nu^i(\nu,T,p) \tag{10}$$

Journal of Physics: Conference Series **275** (2011) 012009 doi:10.1088/1742-6596/275/1/012009

where r_0 is the electron radius, c the velocity of light, $f_{mn}^{a,i}$ the absorption oscillator strength of the spectral line for electronic transitions between the m and n energy states. $N_n^a(T,p)$ is the population density of the lower electronic energy state and $P_\nu^i(\nu,T,p)$ is the normalized line shape.

To calculate the absorption coefficient $\kappa_{a,i}^{bb}$ of the i-th spectral line it is necessary to determine its half-width δ, spectral shift d, and the absorption oscillator strength $f_{mn}^{a,i}$. The values of absorption oscillator strength are tabulated for a lot of spectral lines of atoms and their ions in Kurucz Atomic Line Database [11]. The line shapes in our calculations are given by a convolution of Doppler and Lorentz profiles, resulting in simplified Voigt profiles. The fine multiplet structure and the possible overlap of lines have also been taken into account. For each spectral line we have calculated parameters defining their half-widths and shifts. Semi-empirical formulas given in [9] were used. We have taken into consideration more than 15 000 spectral lines.

3.3. Molecular absorption – diatomic molecules

Changing the orbit of an electron requires a relatively large amount of energy resulting in absorption-emission lines at short wavelengths between the ultraviolet and the near –infrared (between $3\text{x}10^{15}$ s^{-1} and $2\text{x}10^{14}$ s^{-1}). Changes in vibrational energy level require somewhat less energy, so that their spectral lines are found in the infrared (between $2\text{x}10^{14}$ s^{-1} and 10^{13} s^{-1}), while changes in rotational energy levels call for the least amount of energy and rotational lines are found in the far infrared (beyond 10^{13} s^{-1}). Changes in vibrational energy levels are often accompanied by rotational transitions, leading to closely spaced groups of spectral lines that may partly overlap and lead to so-called vibration-rotational bands. Thus, the molecular spectrum is much more complicated then that of monatomic gases.

For electronic transitions no selection rules for vibration quantum numbers exist, i.e. transitions between arbitrary vibration levels of the upper and lower electronic states are possible. The theoretical background for the vibrational structure analysis is the Franck-Condon principle [12] that makes it possible to predict the distribution of vibrational band intensities. Relative line intensities in the electronic-vibrational spectra are determined by the Franck-Condon factors

$$\left[\int \psi_{v'}\psi_{v''}\, d\tau\right]^2 \equiv q_{v'v''} \,. \tag{11}$$

Here, $\psi_{v''}$ and $\psi_{v'}$ are the vibrational wave functions of the lower and higher electronic states, respectively. The bound-bound absorption coefficient for a transition between two vibrational levels v'', v' of two electronic states can be expressed in the form

$$\kappa^{v''v'} = \pi\, r_0 c N_{v''} f_{v''v'} \tag{12}$$

where r_0 denotes the electron radius, c the speed of light, $N_{v''}$ the number density of the lower vibrational state, and $f_{v''v'}$ the band oscillator strength defined by [13]

$$f_{v''v'} = \frac{\nu_{v''v'}}{3R_\infty c}\frac{\overline{R}_e^2}{g''} q_{v''v'} \tag{13}$$

where R_∞ is the Rydberg constant, g'' the weighting factor of the lower level, $\nu_{v''v'}$ the frequency at which the transition occurs, and $\overline{R}_e^2$ gives the intensity distribution in electronic structure. For approximate calculation of radiative properties, it is useful to define the spectral absorption coefficient for the band system averaged over the rotational spectrum, and also partially smeared out over the vibrational structure

$$\overline{\kappa}_\nu = N\sigma(\nu,T) = N\pi\, r_0 \frac{\nu}{3R_\infty}\frac{\overline{R}_e^2}{g''}\frac{1}{\Delta\nu}\sum_{v''v'} q_{v''v'} w_{v''} \,. \tag{14}$$

Here, N is the molecular number density, ν the radiation frequency of absorbed photon, and $w_{v''}$ denotes the Boltzman probability of the occupation of the lower vibrational level. The sum of Franck-Condon factors is performed for all transitions $v'' \rightarrow v'$ with

$$\nu_{v''v'} \in \left\langle \nu - \frac{\Delta\nu}{2}, \nu + \frac{\Delta\nu}{2} \right\rangle$$

where $\Delta\nu$ is interval of averaging around ν. The Franck-Condon factors are tabulated for many molecular transitions in [14], or they can be calculated by computer code [15]. Molecular photo-absorption was taken into account for selected electronic transitions of the following diatomic molecules and electronic transitions: O_2, O_2^+, N_2, N_2^+, NO, NO^+, CO, CO^+, and CN.

Calculated total absorption coefficients of radiation as a function of radiation frequency for plasma temperatures of 6 000 K and 20 000 K are plotted in figure 1. Molecular species (electronic transitions) contribute to the total absorption coefficients slightly at lower temperatures below 10 000 K at radiation frequencies up to 2.5×10^{15} s^{-1}. Due to the frequency interval under consideration, radiation transitions between rotational levels were not taken into account. With increasing temperature the contribution of discrete radiation to the absorption coefficient increases.

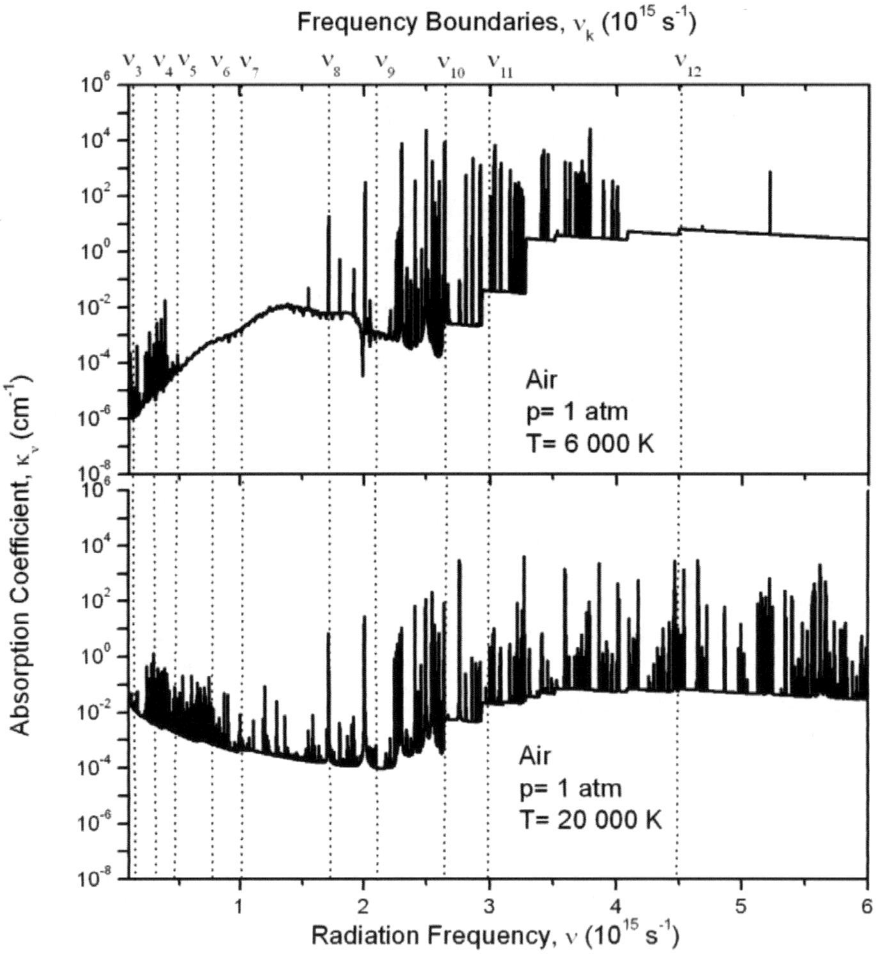

Figure 1. The total absorption coefficient as a function of frequency for thermal air plasmas at temperature of 6 000 K and 20 000 K. Doted lines represent boundaries between specific frequency groups for non-uniform splitting (see table 1).

4. Multigroup method – mean absorption coefficients

One of the methods for handling the frequency dependence in radiation transfer is the multigroup method [16]. One assigns a given photon to one of G frequency groups, and all photons within a given group are treated the same from the point of view of the absorption properties of the medium, the absorption coefficient for given frequency group k is supposed to be constant with certain average value

$$\kappa_\nu(\vec{r},\nu,T)=\kappa_k(\vec{r},T), \quad \nu_k \le \nu \le \nu_{k+1}, \quad k=1,\dots,G. \tag{15}$$

The values of total intensity are then given by

$$I(\vec{r},\vec{\Omega})=\sum_{k=1}^{G} I_k(\vec{r},\vec{\Omega}) \quad . \quad I_k(\vec{r},\vec{\Omega})=\int_{\nu_k}^{\nu_{k+1}} I_\nu(\vec{r},\nu,\vec{\Omega})d\nu \tag{16}$$

The equation of transfer for the given frequency group can be treated as equation for grey medium:

$$\vec{\Omega}\cdot \mathrm{grad}\, I_k(\vec{r},\vec{\Omega})=\bar{\kappa}_k\left(b_k \frac{\sigma}{\pi}T^4 - I_k(\vec{r},\vec{\Omega})\right) \quad , \quad 1\le k\le \mathrm{G} \tag{17}$$

where σ is the Stefan-Boltzmann constant, b_k the fraction of $\frac{\sigma}{\pi}T^4$ lying within the k-th group, i.e.

$$b_k = \frac{\int_{\nu_{k-1}}^{\nu_k} B(\nu)d\nu}{\int_0^\infty B(\nu)d\nu} = \frac{\int_{\nu_{k-1}}^{\nu_k} B(\nu)d\nu}{\frac{\sigma}{\pi}T^4}. \tag{18}$$

For the multigroup method to be useful, one must be able to compute or estimate the mean values of absorption coefficients. Generally, $\bar{\kappa}_k$ is taken as either Rosseland or Planck mean.

4.1. The Rosseland mean

The Rosseland mean, also called the mean free path of radiation is appropriate when the system approaches equilibrium (almost all radiation is reabsorbed). The Rosseland mean absorption coefficient is given by

$$\overline{\kappa_R^{-1}} = \frac{\int_{\nu_{k-1}}^{\nu_k} \kappa_\nu^{-1}\frac{dB(\nu,T)}{dT}d\nu}{\int_{\nu_{k-1}}^{\nu_k} \frac{dB(\nu,T)}{dT}d\nu}. \tag{19}$$

4.2. The Planck mean

The Planck mean is appropriate in the case of optically thin system. The Planck mean absorption coefficient is given by

$$\bar{\kappa}_P = \frac{\int_{\nu_{k-1}}^{\nu_k} \kappa_\nu B(\nu,T)d\nu}{\int_{\nu_{k-1}}^{\nu_k} B(\nu,T)d\nu}. \tag{20}$$

The values of mean absorption coefficients depend on the choice of the frequency interval splitting. As example, results for two different splitting of frequency interval $(0.01 - 6)\times10^{15}$ s^{-1} are presented. Boundaries between 12 specific frequency groups of a non-uniform splitting are given in table 1 and shown in figure 1, boundaries between 7 frequency groups of a quasi-uniform splitting are given in table 2. Values of Planck mean absorption coefficients for two different splitting (non-uniform and quasi-uniform) are shown in figures 2 and 3, respectively.

Table 1. Boundaries between specific frequency groups ($\nu_k\times10^{15}$ s^{-1}), non-uniform splitting.

ν_1	ν_2	ν_3	ν_4	ν_5	ν_6	ν_7	ν_8	ν_9	ν_{10}	ν_{11}	ν_{12}	ν_{13}
0.01	0.03	0.057	0.296	0.386	0.746	0.986	1.71	2.098	2.64	2.997	4.49	6.00

Table 2. Boundaries between specific frequency groups ($\nu_k'\times10^{15}$ s^{-1}), quasi-uniform splitting.

ν_1'	ν_2'	ν_3'	ν_4'	ν_5'	ν_6'	ν_7'	ν_8'
0.01	0.1	1.00	2.00	3.00	4.00	5.00	6.00

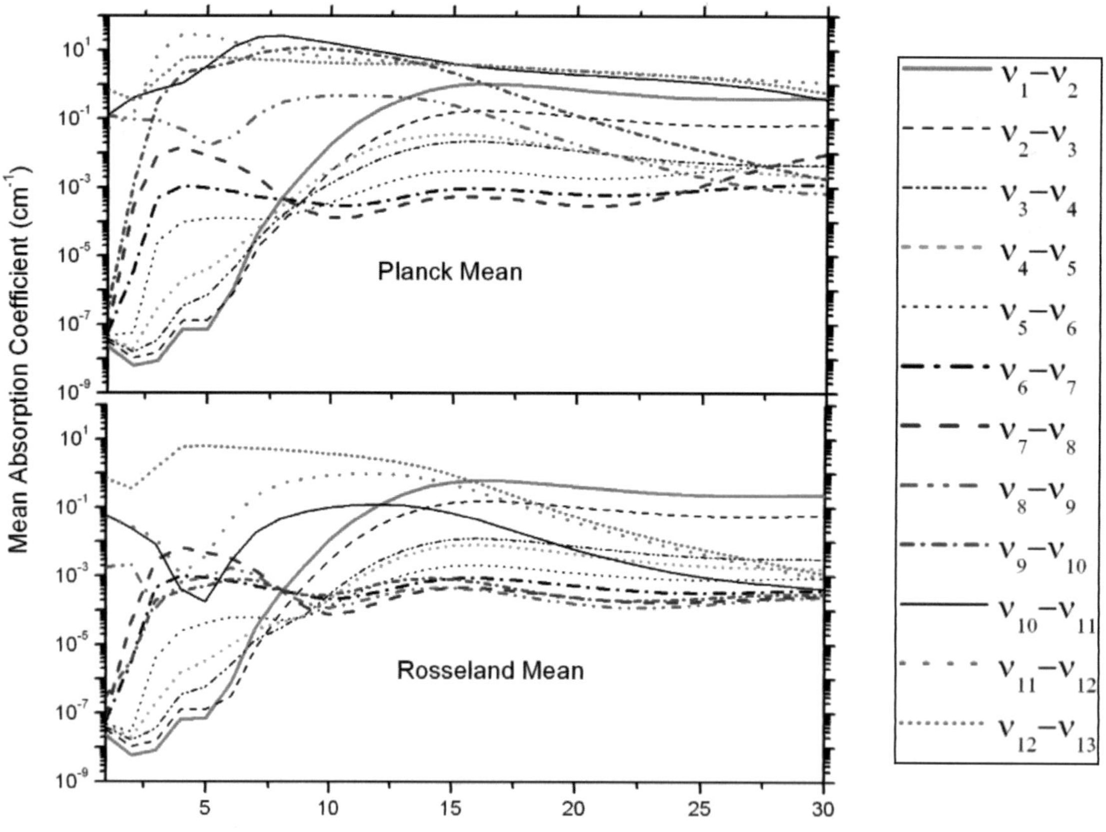

Figure 2. Mean absorption coefficients for 12 specific frequency groups, non-uniform splitting.

Journal of Physics: Conference Series **275** (2011) 012009 doi:10.1088/1742-6596/275/1/012009

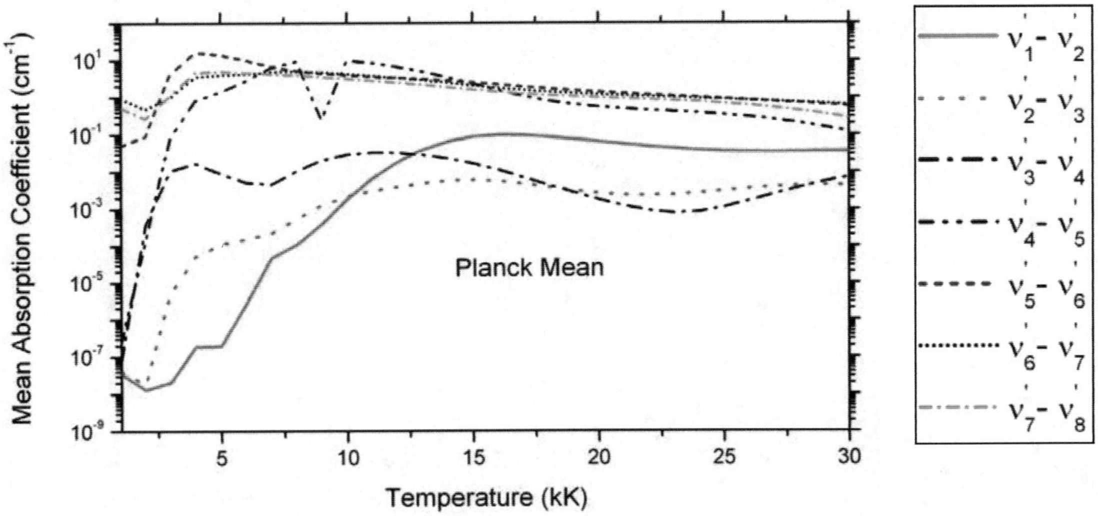

Figure 3. Planck mean absorption coefficient for 7 specific frequency groups, the quasi-uniform splitting.

The Rosseland and Planck mean absorption coefficients can differ by an order of magnitude or more, as can be seen in figure 4 where comparisons of Planck and Rosseland means for two different frequency groups of the non-uniform interval splitting are shown. For the group ν_7 - ν_8 = (0.986 - 1.71)×10^{15} s^{-1} both means differ only slightly up to the temperature 25 000 K. Great difference for the group ν_9 - ν_{10} = (2.098 - 2.64)×10^{15} s^{-1} appears due to strong discrete radiation in this frequency group which makes the Planck mean by several orders of magnitude greater than Rosseland mean. Comparisons of Planck and Rosseland means for all 12 groups of the non-uniform interval splitting are shown in figure 2.

The use of Rosseland and Planck mean absorption coefficients is only strictly appropriate in limiting circumstances (absorption dominated, or emission dominated system, respectively). As can be seen in figures 2 and 4, these two means can differ very strongly, and thus the results of the multigroup method can vary widely depending upon which mean is used. In reality, neither mean is correct in general. Anyway, the splitting of the whole frequency interval has to be made according to the frequency dependence of the absorption coefficients. The simplest procedure is then to use the Planck mean for frequency groups with low values of absorption coefficients; for groups with high values of absorption coefficients the Rosseland mean is more suitable. More accurate results brings the new multigroup method [16] which involves generalization of the Planck and Rosseland means which occur together in the multigroup equations.

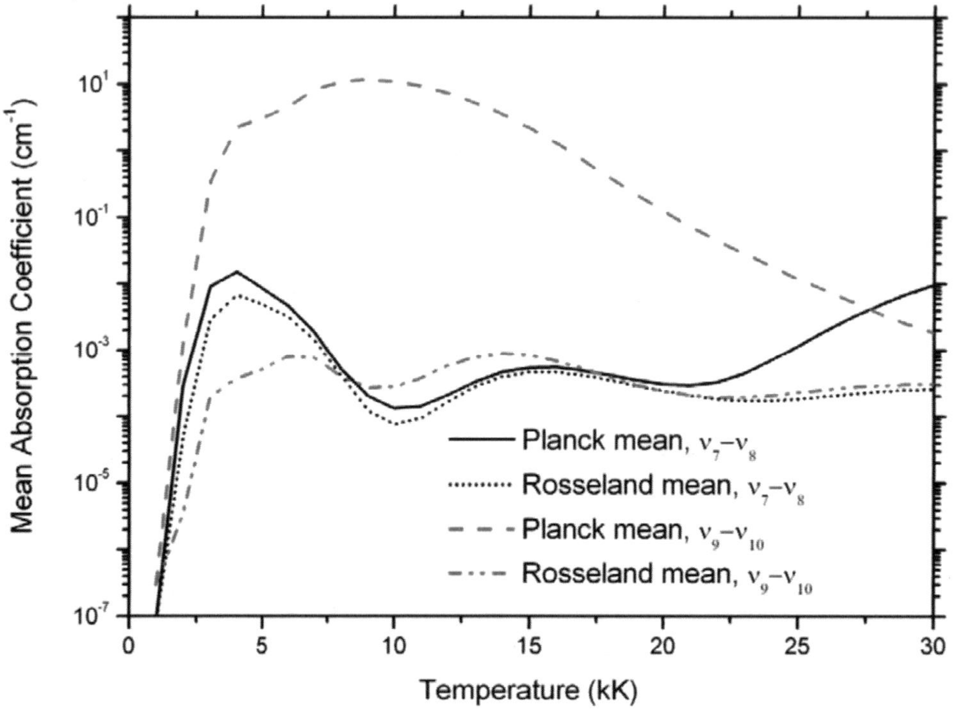

Figure 4. Planck and Rosseland mean for two different specific frequency groups of the non-uniform splitting, ν_7 - ν_8 and ν_9 - ν_{10}.

5. Conclusions

Calculations of the absorption coefficients for thermal plasma of air have been performed as a function of temperature in frequency interval (10^{13} - 6×10^{15}) s^{-1} at the pressure of 1 atm. Both Planck and Rosseland means have been computed for different frequency interval splitting procedures. Results for 12 groups of the non-uniform splitting and 7 groups of the quasi-uniform splitting have been presented. The use of Planck or Rosseland means in the multigroup method of solution of the equation of radiation transfer depends on values of absorption coefficients.

Acknowledgements

This work has been supported by the Czech Science Foundation under project No. GD102/09/H074 and from Ministry of Education, Youth and Sports under project No. MSM0021630503.

References

[1] Lowke J J 1974 *JQSRT* **14** 111-122
[2] Naghizadeh-Kashani Y, Cressault Y and Gleizes A 2002 *J.Phys. D: Appl. Phys.* **35** 2925-2934
[3] Aubrecht V and Bartlova M 2009 *Plasma Chem Plasma Process* **29** 131-147
[4] Aubrecht V and Lowke J J 1994 *J.Phys. D: Appl. Phys.* **27** 2066-2073
[5] Raynal G and Gleizes A 1995 *Plasma Sources Sci. Technol.* **4** 152-160
[6] Coufal O, Sezemsky P and Zivny O 2005 *J.Phys. D: Appl. Phys.* **38** 1265-1274
[7] Verner D A, Ferland G J and Korista K T 1996 *Astrophysical Journal* **465** 487/-492

[8] Burges A and Seaton M 1958 *Rev.Mod.Phys.* **30** 992/-1003
[9] Liebermann R W and Lowke J J 1976 *JQSRT* **16** 253-264
[10] NIST scientific and technical database [http://physics.nist.gov/]
[11] Kurucz Atomic Database [http://www.cfa.harvard.edu/amp/ampdata/kurucz23/sekur.html]
[12] Herzberg G 1950 *Molecular Spectra and Molecular Structure. I. Spectra of Diatomic Molecules* (New York: D. Van Nostrand Co)
[13] Mnacakanjan A Ch 1968 *TVT* **6** 236-241
[14] Kuzmenko N E, Kuznecova L A and Kuzjakov J J 1994 *Faktory Franka-Kondona dvuchatomnych molekul* (Moskva: Izd. Moskovskogo universiteta)
[15] Stakhursky V L [http://www.chemistry.ohio-state.edu/vstakhur]
[16] Pomraning G C 1973 *The Equations of Radiation Hydrodynamics* (New York: Dover Publications)

Journal of Physics: Conference Series **275** (2011) 012010 doi:10.1088/1742-6596/275/1/012010

Application of aqueous dispersions of silver nanostructures for treatment of pyoinflammatory diseases with a chronic component

Ph Rutberg[1], V Kolikov[1], V Snetov[1], A Stogov[1], A Moshkin[2], and M Khalilov[2]

[1] Institute for Electrophysics and Electric Power Russian Academy of Sciences, 18 Dvortsovaya nab., St.-Petersburg, 191186, Russia
[2] Oryol State University, Medical Institute, October st. 25, Oryol, 302028, Russia

Abstract.
Bactericidal properties of aqueous dispersions of oxide silver nanostructures (ADSN) produced by means of pulsed electric discharges (PED) in water can use in surgery for treatment of upper purulent wounds with a chronic component. The patients with such wounds are of large number and differ on etiology of diseases but their mutual feature is long treatment without marked positive changes. Thus long application of antibiotics leads to abnormality of immune processes and antibacterial resistance of microbial flora. Moreover, local antiseptics are frequently toxic and one can oppress processes of reparation in a wound. The investigation is addressed to finding out the opportunity of usage of an ADSN for treatment of purulent wounds with a chronic component and comparison of its efficiency with the sodium hypochlorite. At investigation, the ADSN formed at PED of 5 - 10 μs duration, with highest share of "small" (hydrodynamic diameter $\leq$ 100 nm) nanostructures and greatest surface electric charge we used. It was found that the usage of ADSN during the first 5 days characterized by high active reparative processes with their maximum at 3rd – 4th days and subsequent moderate further healing. At local use of ADSN, there were no cellular atypia and preternatural representations about inflammatory reactions. It is possible to assume that usage of ADSN will allow in prospect to correct the practice of out-patient therapy of chronic and slow pyoinflammatory diseases.

1. Introduction

Surgical infection is one of the priorities in clinical medicine due to high number of morbidity and significant difficulties in treatment. Surgical infection reaches 35% in total surgical morbidity and it takes place in the form of acute and chronic purulent processes.

ADSN produced by means of PED in water have high biological activity and bactericidity. However, to date, many issues of action ADSN on purulent wounds are poorly understood, so they are vital so far.

The study is addressed to determination the possibility of ADSN for development of a method for purulent wounds treatment.

2. Materials and procedure

One of effective ways of oxide metal nanostructures manufacture is processing of water by PED. In this case, ADSN formed due to electrodes erosion by PED contains both nanostructures of electrode metal or its oxides and metal's ions [4, 7]. For determination of ADSN parameters, the techniques of dynamic quasi-elastic scattering and high effective ionic capillary electrophoresis were used.

Distribution of a share of "small" nanostructures subfractions (≤ 100 nm) and a surface electric charge of nanostructures depending on pulse duration by laser correlation spectrometer "LCS-03" and the device capillary electrophoresis "NANOPHORE 01" had been determined (Figure 1).

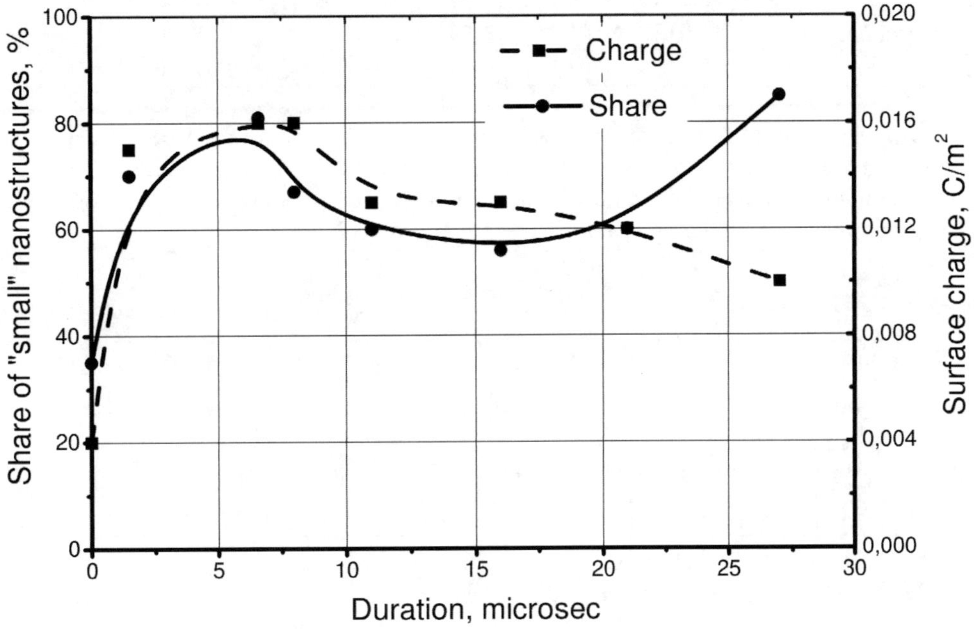

Figure 1. Share and charge of "small" (<100 nm) nanostructures.

For the clinical researches, the ADSN with maximal share of "small" nanostructures and maximal surface electric charge has been chosen. Such nanostructures at pulse duration of 5 - 10 μs are formed. The parameters of electrodischarge unit were next: voltage - 60 kV, pulsed current - 40 A, pulse duration 5 - 20 μs, and pulse frequency - 50 Hz.

Clinical research on 133 rats of both sexes of line "VISTAR" of weight 170 ± 30 g was carried out. The animals were divided in three groups: one control group of 31 animals and two test groups of 51 animals each.

For treatment of wounds, the ADSN of alloy (~ 70 % of silver and ~ 30 % of copper) in 2 months after its preparation and solution of sodium hypochlorite of 600 mg/l were used.

Purulent wounds of initial area of 150 ± 50 mm^2 by means of sewing in inter shoulder-blade areas of a gauze napkin impregnated with a solution of low viral strain of *Staphylococcus aureus* were made. Lancing and surgical sanitation of an abscess on 7th day after infecting were carried out. Treatment procedures at using of ADSN of 4.5 mg/l and solution of sodium hypochlorite of 600 mg/l in test groups of animals in one time per day during 14 days were performed.

For a comprehensive assessment of wound healing process, the methods of planimetric, bacteriological, histological examination of the wounds, which were carried out on the 3rd, 5th, 7th, 10th, 14th days were used.

3. Results and discussion

Bacteriological investigation of purulent wounds included a study of the qualitative composition of microbial pathogens and quantitative count of microorganisms.

Determination of quantitative composition of microflora in wounds per 1 g of tissue is shown in table 1.

Journal of Physics: Conference Series **275** (2011) 012010 doi:10.1088/1742-6596/275/1/012010

Table 1. Number of microorganisms on 1g of tissue in experimental animals.

Groups of animals	1st day (CFU).	3rd day (CFU).	5th day (CFU).	7th day (CFU).
Control	$(3.5\pm1.6)\times10^7$	$(3.9\pm1.9)\times10^7$	$(3.8\pm1.4)\times10^7$	$(3.7\pm1.4)\times10^7$
Sodium hypochlorite	$(3.9\pm1.7)\times10^7$	$(5.3\pm1.6)\times10^5$	$(3.8\pm0.5)\times10^4$	No growth
ADSN	$(3.7\pm1.5)\times10^7$	$(6.5\pm1.1)\times10^5$	$(3.9\pm0.6)\times10^4$	No growth

Thus, the use of sodium hypochlorite solution and ADSN make the effective elimination of the pathogens from the wounds, which was not observed in control group of animals.

Morphodynamic changes of wound healing process at the application of ADSN and sodium hypochlorite compared with the control are presented in table 2.

Table 2. Morphodynamics of wound healing process.

Clinical signs	Groups of animals		
	Control (days)	Sodium hypochlorite (days)	ADSN (days)
Period of wound cleansing	9.1±0.1	4.5±0.2	4.3±0.2
Period of appearance of granulations	8.3±0.2	4.3±0.2	4.2±0.2
Period of wound granulations filling	13.0±0.3	6.0±0.1	5.9±0.2
Beginning of edge epithelialization	10.6±0.2	6.3±0.2	6.3±0.1

The data analysis revealed reducing wound healing time for experimental series using sodium hypochlorite and ADSN for the purulent wound treatment.

Application of sodium hypochlorite and ADSN for the purulent wound treatment led to changes in the area of wounds in experimental animals. The data presented in table 3.

Table 3. The area of wounds in experimental animals during treatment.

Days	Groups of animals		
	Control (mm^2)	Sodium hypochlorite (mm^2)	ADSN (mm^2)
1	40.2±5.1	400.1±3.1	400.2±6.2
3	390.1±1.9	355.2±2.5	333.5±5.1
5	366.4±0.7	238.5±6.3	219.4±4.7
7	355.7±8	182.3±13.5	165.7±9.1
10	334.2±1.1	154.4±10.4	138.9±11.3
14	275.8±0.55	70.0±5.3	52.3±2.3

It should be noted that there is diurnal decrease of wound areas in experimental animals due to daily application of sodium hypochlorite and ADSN. The data presented in table 4.

Table 4. Daily decrease of wound areas in experimental animals.

Days	Groups of animals		
	Control (%)	Sodium hypochlorite (%)	ADSN (%)
3	2.5±0.1	11.2±0.2	16.6±0/2
5	6.0±0.1	32.9±0.1	34.2±0.1
7	2.9±0.1	23.4±0.1	24.4±0.1
10	6.0±0.2	15.3±0.1	16.1±0.1
14	17.4±0.3	54.6±0.1	62.3±0.1

Analysis of the data on the daily decrease in wound areas shows a positive impact of sodium hypochlorite and ADSN on the purulent wound. It should be noted that more effective impact on wound healing has ADSN. In the control group of animals the dynamics of decreasing the surface area of wounds was significantly smaller. Total reduction in the area of wounds for 14th day is presented in table 5.

Table 5. Total reduction in the area of wounds for 14th day.

Groups of animals		
Control (%)	Sodium hypochlorite (%)	ADSN (%)
31.08±0.2	82.5±0.1	86.9±0.1

Data of table 5 shows that the greatest common factor of decreasing the wound areas for 14 days stated in the test group with the use of ADSN.

Changes of wound areas in mm/day during treatment are presented in table 6.

Table 6. Terms of wound healing in experimental animals.

Days	Groups of animals		
	Control (mm^2/day)	Sodium hypochlorite (mm^2/day)	ADSN (mm^2/day)
3	10.1±1.2	44.9±3.5	66.7±2.5
5	23.7±2.7	117.0±5.6	114.1±5.2
7	10.7±1.1	55.9±6.7	53.7±5.4
10	21.5±1.2	27.9±4.5	26.8±3.7
14	58.4±5.7	84.4±6.5	86.6±4.9

The data of table 6 shows that the shortest time of healing was observed with the use of ADSN. This period was slightly shorter than that in case of sodium hypochlorite solution application.

The analysis of planimetric data and the healing time shows high activity of wound healing at using of ADSN, which exceed that of sodium hypochlorite solution.

Thus, the findings of our investigation of wound healing in animals with purulent wounds indicated a reasonably effectiveness of ADSN, which not lower than sodium hypochlorite solution.

Our study has shown that ADSN with local application for the treatment of purulent wounds can be recommended as a means of comprehensive prevention and healing of wound infections because of its antiseptic properties.

Current practice of treatment of chronic purulent wounds on the background of the damaged tissue metabolism and local pathomorphism shows that there is no reason to rely on their quick healing, and therefore, the aim of treatment is to prevent complications.

In addition, the reactivation of an existing microflora when adapting it to antibiotics and the accession of additional associations of pathogens in the nutrient medium of wounds result to considerable adverse effects. Antibiotic resistance is becoming every year more and more pressing problem. Traditionally used antiseptics can inhibit regeneration in the wound and sometimes they lead to total intoxication patient with their long-term use in large-affected areas.

The investigation showed that the ADSN have not of systemic toxicity, which may open new possibilities to cure a broad category of patients. Another significant positive quality of ADSN is the relative ease and cheapness of their manufacturing, as well as the fact that their storage does not require special conditions and they retain their antibacterial properties for a long time.

4. Conclusion

For realization of investigation was used ADSN, which has maximal share of “small” nanostructures (<100 nm) with the greatest surface electric charge. They formed at pulse duration of 5 - 10 μs.

It was founded that bacterial contamination of purulent wounds disappears for the 7th day in the local use of ADSN. The acceleration of wound cleansing and a positive healing morphodynamics with the earlier boundary epithelization was noted. As a result of local application of ADSN enhances the appearance of early signs of purulent wounds healing in comparison with sodium hypochlorite solution.

The attributes of cellular atypia and preternatural representations about inflammatory reactions are not revealed at local use of ADSN.

The positive results of local use of ADSN in the study allow us to recommend this method for use in clinical practice.

Acknowledgements

Work was supported in part by the Russian Foundation for basic research, projects №№ 08-08-00477-a, 10-08-00239-a.

References

[1] Ljapunov A N, Datsenko B M, and Mohert N A 1995 *The problem of medicinal therapy* Theory and practice of local treatment of purulent wounds Kiev p.132-196 (In Russian)

[2] Kolikov V A, Kurochkin V E, Panina L K, and Rutberg F G 2005 *Doklady Biological Sciences,* Pulse electric discharges and prolonged microbial resistance of water **403** p.279-281

[3] Kolikov V A, Kurochkin V E, Panina L K, Rutberg A F, Rutberg F G, Snetov V N, and Stogov A J 2007 *Technical Physics* Prolonged microbial resistance of water treated by a pulsed electric discharges **52**, no.2 p.263-70

[4] Fedorov D N, Ivashkin A N, Shinin V V, Vasiljev A V, and Ivanov A A 2002 *Archive of a pathology* Morphological and immunological characteristic conversion processes in long not healing wounds **64** № 1 p.8-11(In Russian)

[5] Rutberg Ph G, Kolikov V A, Kurochkin V E, Panina L K, and Rutberg A Ph 2007 *IEEE Transactions on Plasma Science* Electric Discharges and the Prolonged Microbial Resistance of Water **35** no.4 p.1111-18

[6] Rutberg Ph, Kolikov V, Snetov V, Stogov A, Noskin L, Landa S, and Arutunjan A 2007 *IEEE Pulsed Power and Plasma Science Conference (PPPS-2007)* Pulsed electric discharges in water and oxide nanostructures p.1244-47.

Journal of Physics: Conference Series **275** (2011) 012011 doi:10.1088/1742-6596/275/1/012011

Effects of hydrogen addition in nitrogen atmospheric pressure plasma on its optical and electrical properties and silicon-based deposits compositions

D Debrabandere[1], X Vanden Eynde[1] and F Reniers[2]
[1] CRM, Centre for Research in Metallurgy, Liège, Belgium
[2] Analytical and Interfacial Chemistry, Faculty of Sciences, Université Libre de Bruxelles, Bruxelles, Belgium

E-mail: debrabandere@rdmetal.ulg.ac.be, vdeneynde@rdmetal.ulg.ac.be, freniers@ulb.ac.be

Abstract. Si-based coatings were deposited with a cold plasma jet (Plasmabrush® PB1 from Reinhausen Plasma) at atmospheric pressure with nitrogen as main carrier gas and hexamethyldisilazane (HMDSN) as precursor. Effects of hydrogen addition on the plasma characteristics and the coatings compositions have been evidenced with optical emission spectroscopy (OES), power measurements and XPS in-depth analyses. The intensity evolution of the nitrogen line (at 315.9 nm) with the applied voltage has a sigmoid shape for the pure nitrogen plasma but a quite linear one with hydrogen addition (up to 3%). Based on OES spectra, the presence of the NH specie in the nitrogen-hydrogen plasmas has been evidenced (around 336.0 nm) but not in the pure nitrogen plasmas. Although the plasma power is similar for both gases, the nitrogen atomic concentrations in the films as evidenced by XPS were lower with the nitrogen-hydrogen plasmas than with the pure nitrogen plasmas indicating a chemical effect of the presence of hydrogen in the plasma.

1. Introduction

Barrier coatings are widely used in the industry. In particular, silicon oxide films have a good oxidation resistance. These films can be deposited using different methods: thermal chemical vapor deposition (CVD) [1], plasma enhanced CVD (PECVD) at low pressures [2], PECVD at atmospheric pressure [3-4],... CVD processes require high temperatures. For the PECVD processes the thermal energy is replaced by electrical energy. This allows treating the substrates at lower temperatures [5-6]. Furthermore, atmospheric pressure PECVD exempts from maintaining vacuum in a chamber and allows treating large surface areas in continuous processes [3-4, 7]. While using PECVD for deposition processes, some authors inject the precursor in the afterglow region and place their substrates downstream from the plasma [2- 4, 7]. In this remote PECVD system, the substrate is not damaged by energetic ions.

A lot of precursors are used for the deposition of silicon-based coatings: tetraethoxysilane [1, 3-4], hexamethyldisiloxane [2], hexamethyldisilazane [3, 5-6, 8], tetramethylsilane [9], tetramethyldisiloxane [3], tetramethylcyclotetrasiloxane [3], hexamethylcyclotrisilazane [8], bis(trimethylsilyl)carbodiimide [5],...

In this work, Si-based coatings were deposited on aluminum substrates by a commercial cold plasma jet at atmospheric pressure and by introducing hexamethyldisilazane in the post-discharge. The

Journal of Physics: Conference Series **275** (2011) 012011 doi:10.1088/1742-6596/275/1/012011

effect of hydrogen additions on the plasma characteristics was investigated by optical emission spectroscopy (OES) and power measurements. XPS in-depth analyses were used to characterize the coatings obtained by pure nitrogen and nitrogen-hydrogen mixture plasmas.

2. Experimental details

The cold plasma jet used in this work is the commercial Plasmabrush® PB1 system from Reinhausen Plasma (figure 1). This system consists in a central powered electrode surrounded by a grounded electrode with a dielectric between them. It was operated in a chamber filled with nitrogen at atmospheric pressure. The main plasma gas is nitrogen (with a 20 l/min flow rate) or nitrogen (with the same flow rate) with a 3% volume concentration of hydrogen. The precursor used for the deposition is hexamethyldisilazane (HMDSN 98.5% from ABCR). This liquid precursor (with a 56 g/h flow rate) is vaporized and mixed with a separate nitrogen flow (3 l/min) before being introduced in the plasma post-discharge. A new plasma source head in polyoxymethylene was made for this purpose (figure 2) and was used only for the deposition process (not for the plasma OES and electrical measurements performed without any precursor). Precursor mixture is injected at the end of the new plasma source head (figure 3). The aluminum substrate (cut in a 36 µm thickness Reynolds aluminum foil) is put at a distance of 5 mm from the new plasma source head.

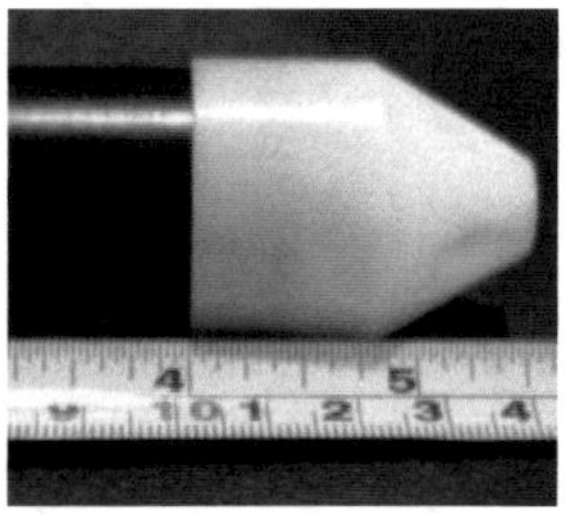

Figure 1. Commercial Plasmabrush system.

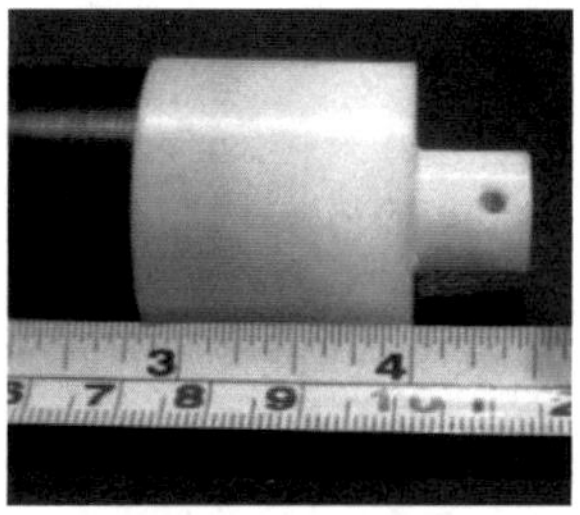

Figure 2. New plasma source head for the precursor injection.

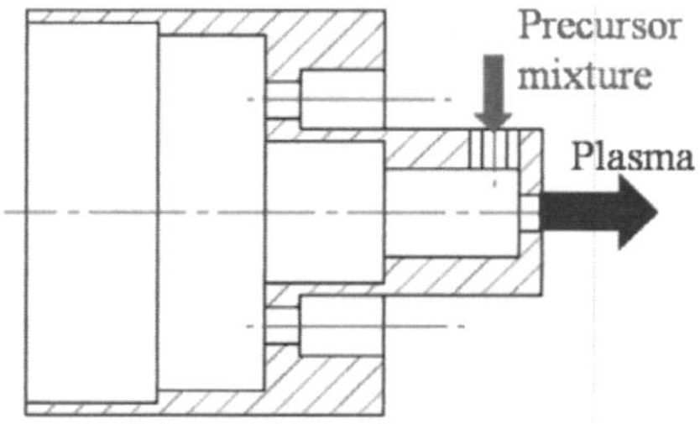

Figure 3. Design of the new plasma source head.

The plasma is characterized by optical emission spectrometry with the HR4000CG spectrometer from Ocean Optics. The wavelength range is 200-1100 nm with a 0.75 nm resolution. The optical fiber is placed vertically below the plasma (looking inside the plasma source). Each spectrum is recorded with a 1 second exposure time (except when specified).

The power delivered to the system is calculated from current and voltage measurements (average over 10 ms). The current sensor is the CWT015B/2.5/700 sensor from Powertek. It is composed of a Rogowski coil and an integrator. Its work principle is described in [10]. Its frequency bandwith ranges from 150 Hz to 2.2 MHz. The coil is wrapped 3 times around the cable placed between the grounded electrode and the ground. The applied voltage is measured using a high voltage probe (TESTEC HVP 15HF, bandwidth: 50 MHz). Both measurements are recorded (one record every 200 ns) with the Tektronix TDS5052 oscilloscope (bandwidth: 500 MHz, maximum sample rate: 2.5 GS/s).

The coatings are analyzed by XPS depth profiles with a PHI-Quantum 2000 spectrometer. The spectrometer is characterized by a monochromatized Al Kα primary X-ray beam and a photoelectron take-off angle of 45 degrees against the sample normal direction. Analysis quantification is based on sensitivity coefficients proposed by the equipment manufacturer. In-depth profiles are obtained by alternating 4 keV-1µA ion sputtering steps and analysis sequences. Sputtering rate is evaluated on iron at 8nm/min.

3. Plasma optical characterization

The evolution of the optical emission spectra of the plasma is studied as a function of the gas composition (N_2 and N_2-H_2 mixture) and the applied voltage varied between 3.0 and 4.5 kV.

The main optical emission lines observed in the optical emission spectra of pure nitrogen (figure 4) and nitrogen-hydrogen (figure 5) plasmas are the nitrogen second positive system bands [11-12]. CN violet system bands (around 385 and 388 nm [11]) are also visible on the pure nitrogen plasma spectra but disappear with hydrogen addition in the plasma. The carbon may come from the atmosphere pollutants or from the plasma source head. Although the main atomic hydrogen lines (at 486.1 and 656.3 nm [11]) were not observed in the nitrogen-hydrogen plasmas spectra, the presence of hydrogen in the plasma can be evidenced by the shoulder to lower wavelengths in the nitrogen band at 337.1 nm (figure 6). This shoulder corresponds with the NH band at 336.0 nm [11].

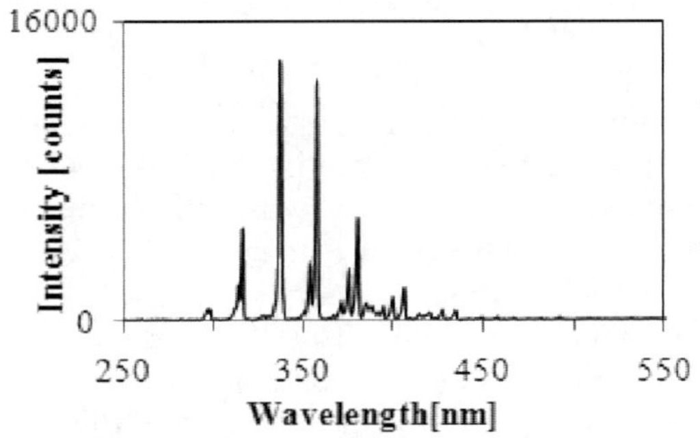

Figure 4. OES spectrum of pure nitrogen plasma at 4.5 kV (exposure time: 0.5 second).

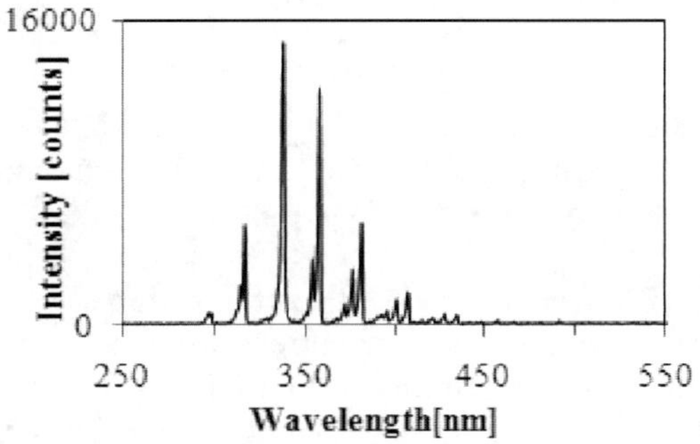

Figure 5. OES spectrum of nitrogen-3% hydrogen plasma at 4.5 kV.

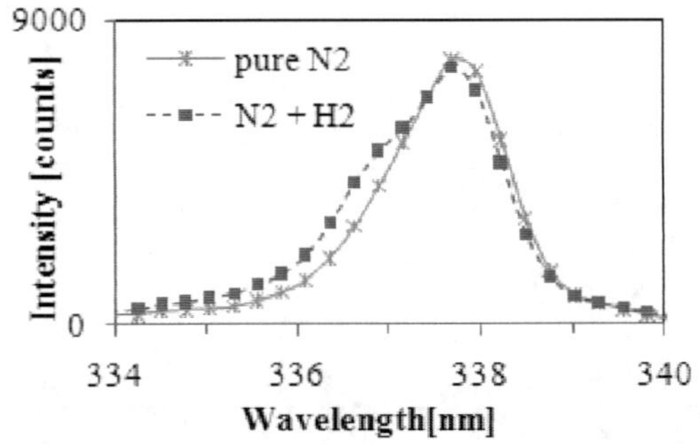

Figure 6. OES spectra of nitrogen with and without hydrogen plasmas at 3.6 kV showing the presence of the NH band for the nitrogen-3% hydrogen plasma.

The figure 7 describes the variation of the nitrogen line (at 315.9 nm) optical emission intensity with the applied voltage for nitrogen with and without hydrogen plasmas. The curve observed for pure nitrogen plasmas (without hydrogen) can be divided in 3 parts. The first part is nearly horizontal at low voltages. The second part is a transition between 3.3 kV and 3.7 kV. The third part shows a nearly linear increase of the emission intensity with the applied voltage. When hydrogen was added in the plasma the transition was not observed. The evolution of the intensity was nearly linear with the applied voltage. This nitrogen line at 315.9 nm is representative of the OES spectrum area called brightness in this paper (figure 8).

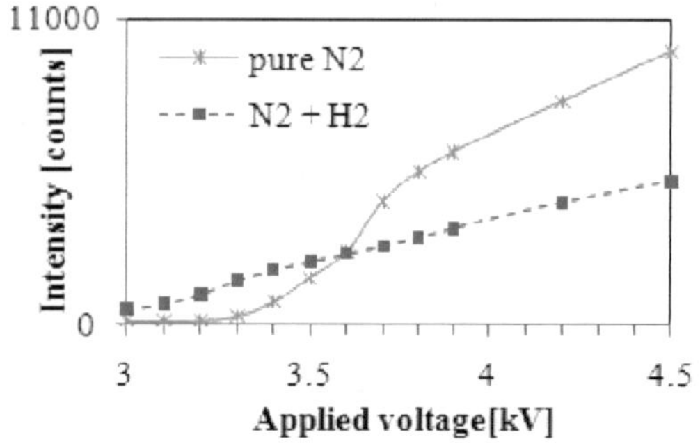

Figure 7. Nitrogen line (315.9 nm) intensity vs. applied voltage for pure nitrogen and nitrogen-3% hydrogen plasmas.

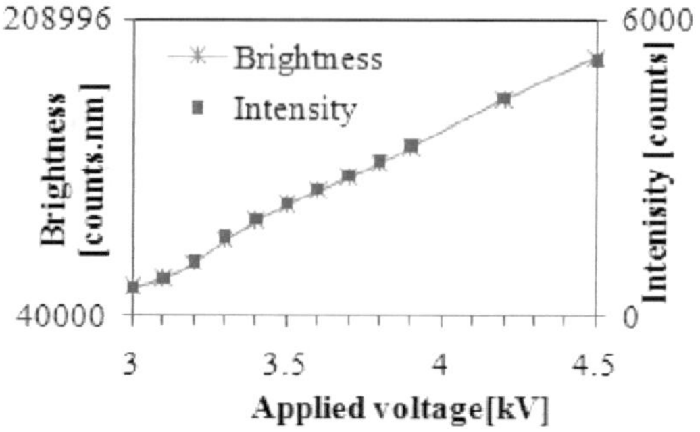

Figure 8. Brightness (spectrum area) and nitrogen line (315.9 nm) intensity vs. applied voltage for nitrogen-3% hydrogen plasma.

4. Plasma electrical characterization

The applied voltage and current are measured for nitrogen and nitrogen-hydrogen plasmas at applied voltages ranging from 3.0 to 4.5 kV. The voltage lags the current by about 85 degrees (figures 9 to 12) which corresponds with a capacitive system such as in a dielectric barrier discharge (DBD). A lot of peaks can be observed on the current signal indicating the discharge is filamentary. When the applied voltage is increased more peaks can be seen. Hydrogen enhances the plasma filamentary behavior and therefore helps the plasma activation at lower voltages.

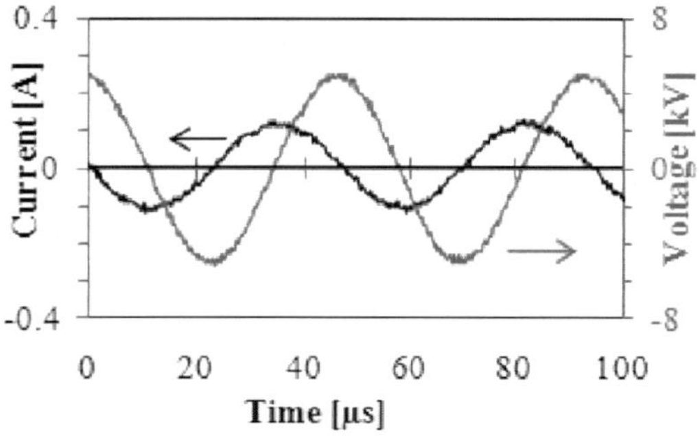

Figure 9. Current and voltage vs. time for pure nitrogen plasma at 3.3 kV.

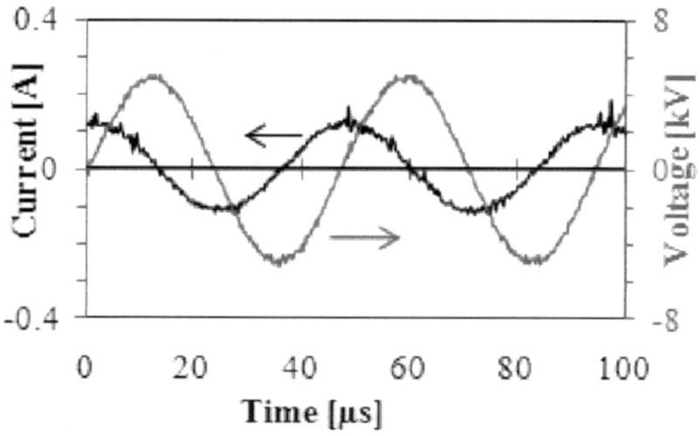

Figure 10. Current and voltage vs. time for nitrogen-3% hydrogen plasma at 3.3 kV.

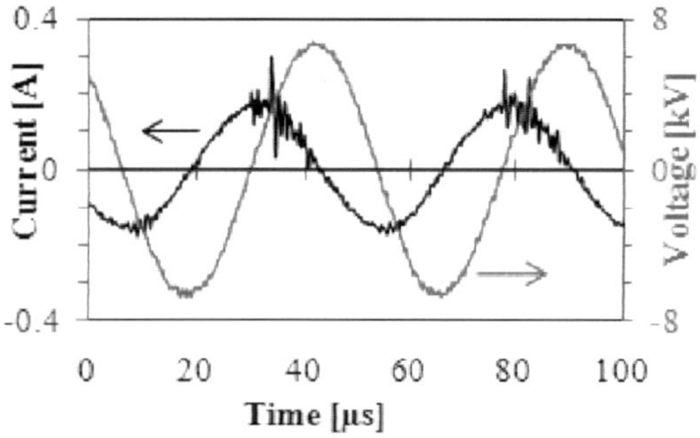

Figure 11. Current and voltage vs. time for pure nitrogen plasma at 4.5 kV.

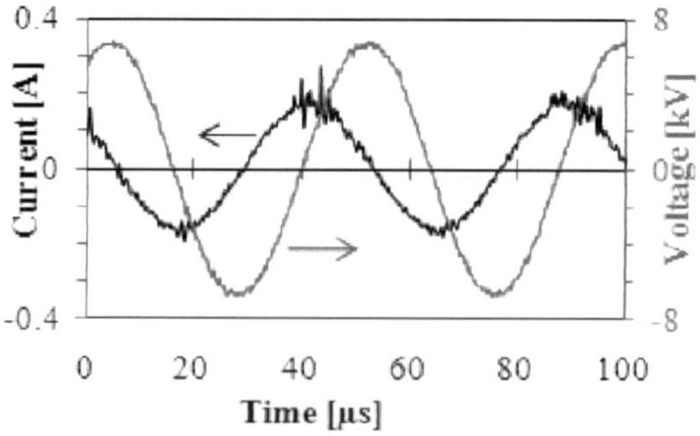

Figure 12. Current and voltage vs. time for nitrogen-3% hydrogen plasma at 4.5 kV.

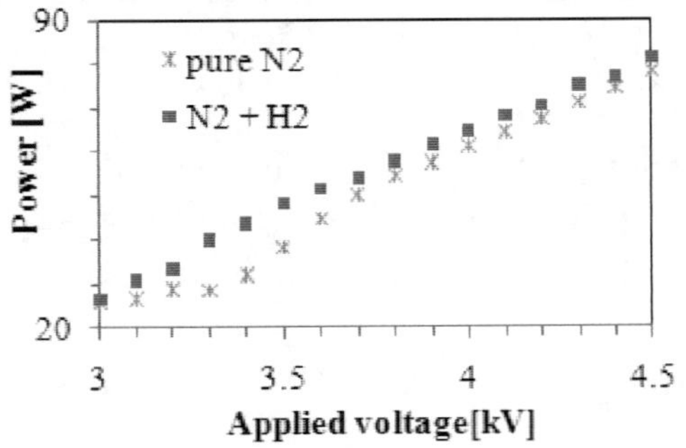

Figure 13. Power vs. applied voltage for pure nitrogen and nitrogen-3% hydrogen plasmas.

As for the nitrogen line optical emission intensity, the evolution of the power was studied as a function of the applied plasma voltage. This evolution is quite linear for nitrogen-hydrogen plasmas but has a sigmoid shape for pure nitrogen plasmas (figure 13). Therefore it partially explains the nitrogen line optical emission intensity evolution with the applied voltage. As the power is in similar range for both gases (except for applied voltages from 3.2 to 3.6 kV), there is also a chemical effect of hydrogen addition in nitrogen plasma.

5. Si-based coatings

Si-based coatings were deposited on aluminum substrates by nitrogen and nitrogen-hydrogen (with a hydrogen volume concentration of 3%) plasmas. The coatings were characterized by XPS depth profiles. The figure 14 shows a XPS depth profile of a coating obtained by pure nitrogen plasma (without hydrogen) with the new plasma source head (figure 2). In that case the film contained not only silicon but also significant levels of oxygen, nitrogen and carbon.

When hydrogen was added in the plasma (figure 15) the carbon and nitrogen atomic concentrations in the film decreased. It has already been observed that the hydrogen presence in the plasma reduces the carbon content in coatings obtained with organosilicon precursors [8].

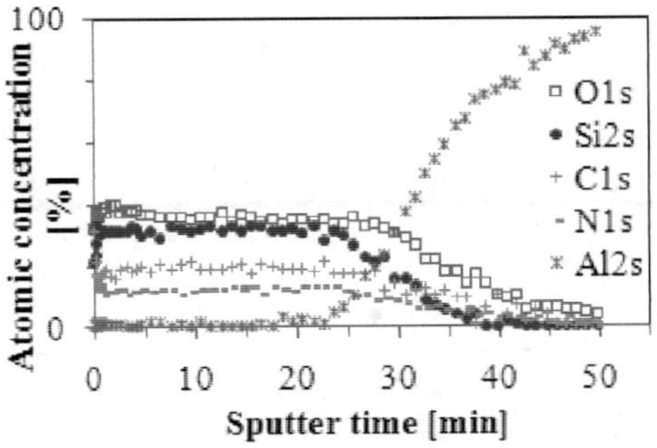

Figure 14. XPS depth profile of a coating deposited on aluminum by pure nitrogen plasma at 4.5 kV during 15 minutes.

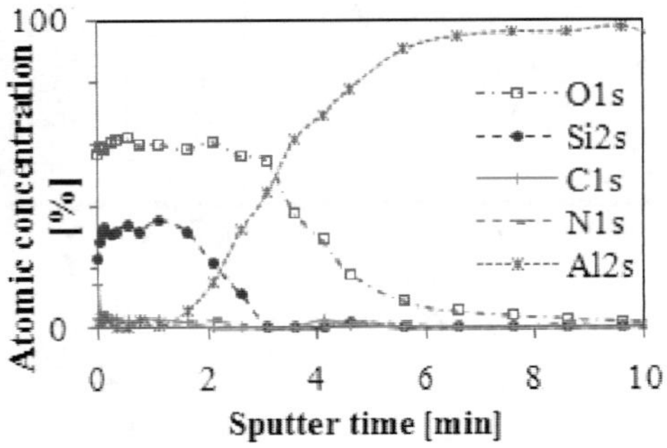

Figure 15. XPS depth profile of a coating deposited on aluminum by nitrogen-3% hydrogen plasma at 4.5 kV during 15 minutes.

6. Conclusions

The addition of a small quantity of hydrogen (3%) in nitrogen plasma has a large influence on the plasma properties as it has been evidenced by optical emission spectroscopy and electrical measurements as well as on the composition of Si-based films deposited with HMDSN.

Based on OES spectra the hydrogen addition (3% volume concentration) in the plasma produces the NH specie. It also "linearizes" the evolutions of the nitrogen line (at 315.9 nm) optical emission intensity (representative of the OES spectrum area) and of the power with the applied voltage whereas these evolutions have a sigmoid shape for pure nitrogen plasmas. As a matter of fact, the total power is similar for both gases when the applied voltage ranges from 3.7 to 4.5 kV and at 3.0 kV but the nitrogen line optical emission intensity is different. Therefore there are chemical and electrical effects of hydrogen addition in nitrogen plasma.

Hydrogen addition in nitrogen atmospheric plasma changes the films composition. The coatings deposited with pure nitrogen plasma are $SiO_xC_yN_z$ but those deposited with nitrogen-hydrogen plasma are close to SiO_2.

Acknowledgments

The authors would like to thank the Walloon Region (Belgian authorities) for financial support in the framework of a FIRST DOCA program.

References

[1] Raupp G B, Schemansky F A and Cale T S 1992 *J. Vac. Sci. Technol.* B **10** 2422-30
[2] Saloum S and Naddaf M 2008 *Vacuum* **82** 50-55
[3] Moravej M and Hicks R F 2005 *Chem. Vap. Deposition* **11** 469-76
[4] Babayan S E, Jeong J Y, Tu V J, Park J, Selwyn G S and Hicks R F 1998 *Plasma Sources Sci. Technol.* **7** 286-88
[5] Probst D et al 2005 *Surf. Coat. Technol.* **200** 355-59
[6] Xiao Z G and Mantei T D 2003 *Surf. Coat. Technol.* **172** 184-88
[7] Nowling G R, Babayan S E, Jankovic V and Hicks R F 2002 *Plasma Sources Sci Technol.* **11** 97-103
[8] Fracassi F, d'Agostino R and Bruno G 1996 *Plasmas and Polymers* **11** 3-16
[9] Thomas L, Maillé L, Badie J M and Ducarroir M 2001 *Surf. Coat. Technol.* **142-44** 314-20
[10] Ward D A and Exon J L T 1993 *IEE Engineering Science and Education Journal* **2** 105-113
[11] Pearse R W B and Gaydon A G 1965 *The Identification of Molecular Spectra* (London: Chapman and Hall)
[12] Navratil Z, Trunec D, Smid R and Lazar L 2006 *Czechoslovak Journal of Physics* **56** B944-51

Journal of Physics: Conference Series **275** (2011) 012012 doi:10.1088/1742-6596/275/1/012012

Investigation of the gas flow effect on an atmospheric pressure RF plasma torch

M Atanasova[1,2], **D Mihailova**[3], **E Carbone**[3], **J van Dijk**[3], **J J A M van der Mullen**[3], **E Benova**[2] and **G Degrez**[1]

[1]Department of Applied Sciences, ULB, 50 av. Fr. Roosevelt – CPI 165/04, Brussels, Belgium

[2]Department for Language Teaching and International Students, SU, 27 Kosta Loulchev St., 1111 Sofia, Bulgaria

[3]Department of Applied Physics, Eindhoven University of Technology, P.O. Box 513, 5600 MB Eindhoven, The Netherlands

E-mail: matanaso@ulb.ac.be

Abstract. A cool atmospheric pressure non-thermal capactively-coupled RF discharge is studied. It is created between two parallel electrodes – a powered one supplied by 13.56 MHz, and a grounded one. The feed gas argon flows via holes between the electrodes where it is ionized. The plasma torch is studied by means of a time dependent two-dimensional fluid model. A simplified kinetic scheme with four active species is considered, namely argon excited atoms (Ar^*), atomic (Ar^+) ions, molecular (Ar_2^+) ions and electrons (e). The plasma dynamics in the space between the electrodes as well as in the extended region behind the grounded electrode is studied. The effect of the gas flow on the plasma is examined. Constriction of the plasma is induced by the field sustaining the discharge due to the sieve-like structure of the electrodes. As a result of the stationary gas flow the filaments extend beyond the electrodes ensuring a flow of active species in the afterglow.

1. Introduction

Atmospheric pressure non-thermal discharges continue to receive formidable interest during the last years. That is because they are easy to operate which consequently leads to numerous practical implementations. Particular interest attracts the radio-frequency atmospheric plasma jet. It is a device that features low gas temperature, that is versatile in selecting the gas chemistry, and adjustable in power input and gas flow rate. These advantages result in various applications such as surface treatment of polymers, surface modification, film coating and etching, developing of nanotechnologies, sterilization, and environmental applications [1–6]. Although there exists advanced experimental and modelling studies a complete picture of the physical and chemical processes determining the performance of this discharge is not obtained, yet. Most of the studies are on the gas breakdown [7–9], the transitions of the discharge from α to γ mode [10,11] and the concentration of radicals in the afterglow [12,13].

The present study focuses on a non-thermal, non-equilibrium, RF argon discharge at atmospheric pressure. It aims to give a better knowledge on the processes and the discharge characteristics inside

Journal of Physics: Conference Series **275** (2011) 012012 doi:10.1088/1742-6596/275/1/012012

the plasma source and thus to help to get an insight into the phenomena within the region which is hardly accessible experimentally. It also provides information on the species densities and their spatial distributions in the effluent region. This is of crucial importance to improve laboratory and industrial applications. The model is used to study the dependence of the properties of the argon plasma on the geometry, the dimensions and the effect of the flow on the discharge composition and kinetics.

2. Model

A hybrid model is constructed that combines the chemical kinetics obtained with a time-dependent, two-dimensional fluid model with the flow provided by a Navier-Stokes solver.

The PLASIMO MD2D sub-model is used to carry out the time-dependent simulations [14]. It features a drift-diffusion description for the charged particles and accounts for the convective motion of the species. The calculation of the convection velocity is carried out by another PLASIMO sub-model. In such a way, a complete description of the constituting particles, the processes, the flow pattern and the discharge characteristics can be accomplished.

2.1. Plasma chemical kinetics

The electron energy distribution function (EEDF) is obtained by solving the Boltzmann equation using the Bolsig+ solver [15]. With the EEDF the electron transport coefficients: mobility μ_e and diffusivity D_e and electron induced reaction rates k_i are calculated. A simplified kinetic scheme with four active species is considered consisting of argon excited atoms (Ar^*), atomic (Ar^+) ions, molecular (Ar_2^+) ions and electrons (e). The energy diagram of argon is presented in figure 1. The considered excited atoms are those of the 4s system treated as a lumped block of levels. This is justified by the high exchange probability of the population of the levels in this level block. The following elementary processes are included in the model: elastic scattering, excitation, direct, stepwise, and Penning ionization, dissociative recombination, molecular ion formation, diffusion and radiation losses with accounting for trapping. The elementary processes taken into account with the corresponding references for the cross sections and rate constants data are listed in Table I. For the ions, the local field approximation was used. The values for the mobility of atomic and molecular ions and excited atoms were adopted from references [16,17]. The diffusion coefficients of all the species were calculated from their mobilities using the Einstein relation.

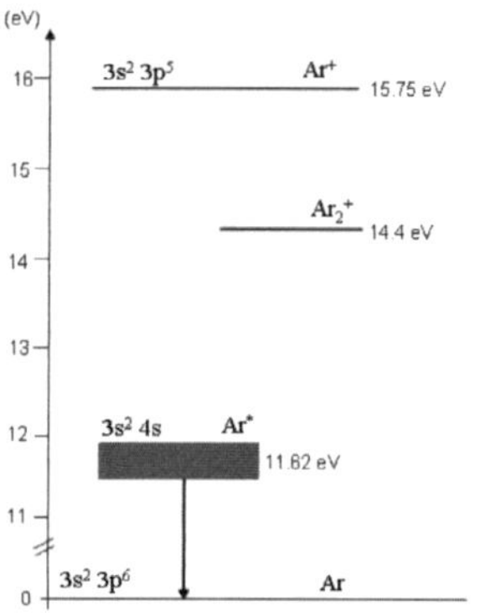

Figure 1. Simplified energy level diagram of the argon atom

Table I. Elementary processes included into the model and corresponding references for the cross sections and rate constants data.

Process	Notation	Reference
Elastic scattering	$e + Ar \rightarrow e + Ar$	[18]
Excitation	$e + Ar \rightarrow e + Ar^*$	[19]
Direct ionization	$e + Ar \rightarrow e + e + Ar^+$	[20]
Stepwise ionization	$e + Ar^* \rightarrow e + e + Ar^+$	[21]
Radiation	$Ar^* \rightarrow Ar + h\nu$	[21]
Penning ionization	$Ar^* + Ar^* \rightarrow Ar^+ + e + Ar$	[22]
Molecular ion formation	$Ar + Ar + Ar^+ \rightarrow Ar + Ar_2^+$	[21,22]
Dissociative recombination	$Ar_2^+ + e \rightarrow Ar + Ar^*$	[23]

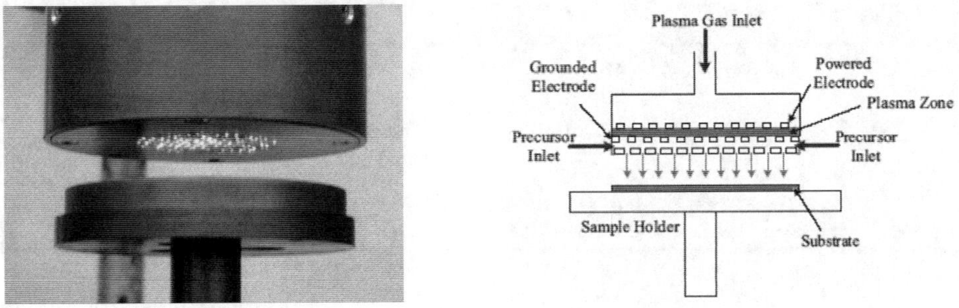

Figure 2. Photo and schematic representation of the atmospheric pressure plasma jet

2.2 Plasma dynamics
The plasma dynamics part couples the transport of the relevant species (electrons, ions, excited atoms), to the production and destruction balances of these species, the electric field and the electron energy. The governing fluid equations are:

- the continuity equations for the excited atoms, ions and electrons;
- the transport equations of the considered particles solved in a drift–diffusion approximation taking a convective velocity field $\mathbf{v}_{conv}$ into account;
- the energy equation for the electrons;
- the Poisson equation for the electric field.

The coupled differential equations are solved by the so-called "modified strongly implicit method" [24] using an extra stabilization method [25].

2.3. Fluid dynamics
To calculate the convection velocity field $\mathbf{v}_{conv}$ a fluid dynamics module is used to calculate the mass-averaged velocity field. For a given mass density field ρ it calculates the velocity $\mathbf{v}$ and the pressure field p. These quantities are determined from the combined solution of the:

- mass continuity equation and the
- momentum balance equation

3. The experimental set-up
Figure 2 gives a photo and a scheme of the atmospheric pressure plasma jet that was considered in this study. It consists of two parallel perforated electrodes separated by a variable gap of 1 to 2 mm. One of the electrodes is driven by a RF between 30 and 100 W at 13.56 MHz, while the other electrode is grounded. Feed gas made of argon (or helium) and up to 0.1 vol% of other gases, flows between the electrodes at a rate of 20 up to 60 L/min. The gas emerges from the holes and impinges on a substrate placed 2 to 15 mm downstream, where cleaning or etching of surface material takes place [12].

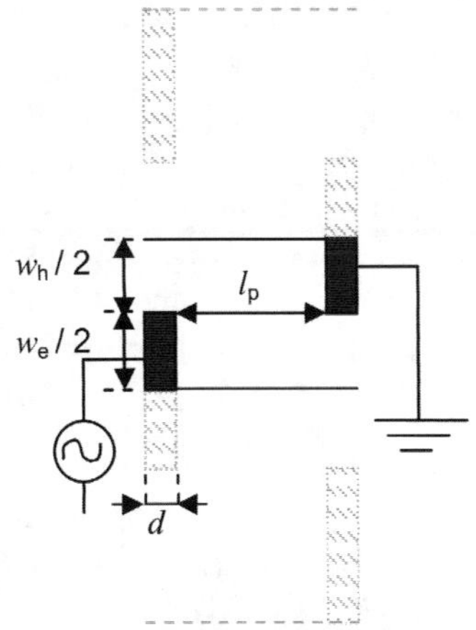

Figure 3. Schematic of a unit cell from the perforated electrodes geometry.

4 Results

4.1. Steady discharge
We model the atmospheric pressure plasma torch using a periodic 2D Cartesian geometry in which the hole-pattern shown in figure 2 is seen as a repetition of identical cells. Such a unit cell is depicted in figure 3. It includes half of a hole and half of an electrode between the two holes. The powered electrode is on the left, the plasma bulk in the middle and a similar electrode pattern on the right. This

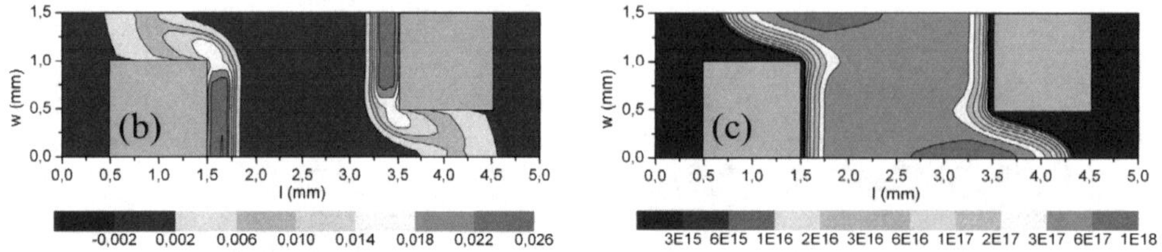

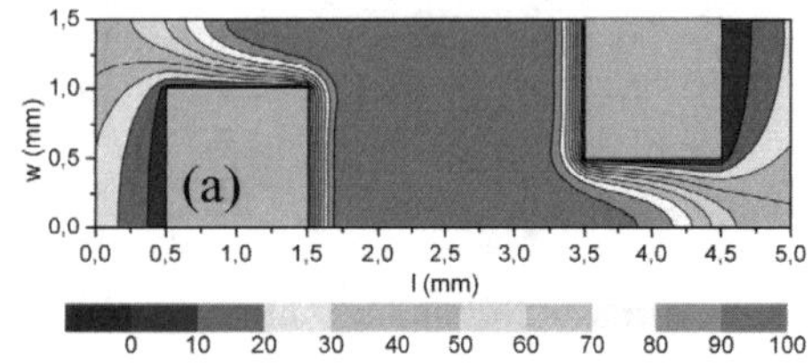

Figure 4. Spatial distribution of the electric field in V (a), volume charge in nC (b) and electron density in m^{-3} (c) between perforated electrodes. The set-up dimensions are: l_p = 2 mm, d = 1 mm, $w_h/2$ = 0.5 mm, $w_e/2$ = 1 mm.

grounded electrode is displaced in vertical direction. On the top and the bottom of the unit cell mirror boundary conditions are applied. In figure 3, l_p denotes the distance between the electrodes, $w_h/2$ the half-width of the hole, $w_e/2$ the half-width of the electrode and d the thickness of the electrode. The calculations are performed for a unit cell with d = 1 mm, w_h = 1 mm, w_e = 2 mm and l_p = 2 mm. For the gas temperature 350 K is taken [13]. The applied voltage is – 300 V.

Searching for an optimal configuration in which we have maximum uniformity and homogeneity of the plasma, we examined first the effect of the set-up dimensions on the discharge characteristics for the non-convective case; i.e. $\mathbf{v}_{conv} = 0$. The spatial distributions of plasma potential, volume charge and electron density are shown in figure 4. These are obtained by averaging over one period of the electric cycle. As shown in figure 4(a), the electric potential has a large gradient close to the electrodes. The sheath regions formed around the electrodes are observable in figure 4(b) as domains with high volume charge. This charge extends also inside the holes. As a result a zigzag particle density pattern and higher density structures perpendicular to the plane of the electrodes and stretched inside the holes are formed (figure 4(c)).

The effect of the distance between the electrodes on the discharge properties is shown in figure 5 where the plasma densities for three l_p- values are plotted. The snapshots are taken at the moment in the RF cycle when a maximum of the negative potential (at the left electrode) is reached. It can be seen in figure 5(a) that for smaller gap-spacing the plasma density is higher but also more unevenly distributed. With the increase of the gap (see also figure 4(c)) the electron density within the so called filaments decreases the plasma becomes more homogeneous. The optimal gap spacing is found to be l_p = 2.5 mm. Higher distances result in a drop of the density of the electrons and thus of all the other species in the discharge. This is shown in figure 5(c).

Another topic of interest is the influence of the inequality between the aspect ratio AR of the left and right electrode. This AR is the ratio of the width of the hole and the electrode, w_h/w_e, Figure 6 shows what happens with the electron density and the volume charge distribution if we keep for the left electrode w_h/w_e fixed to 1/2 while for the right electrode for the ratios w_h/w_e 0.75/2.25 and

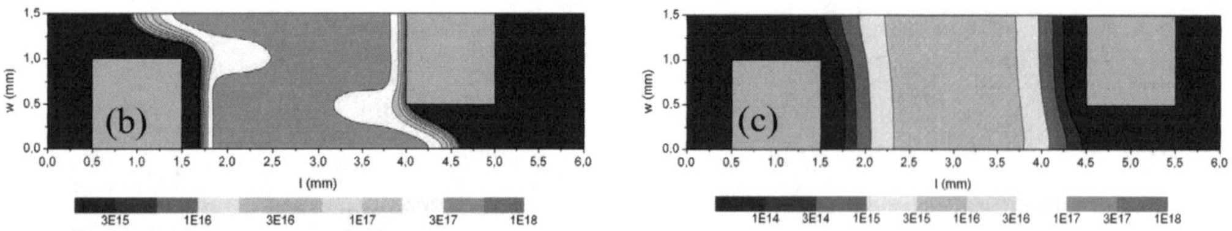

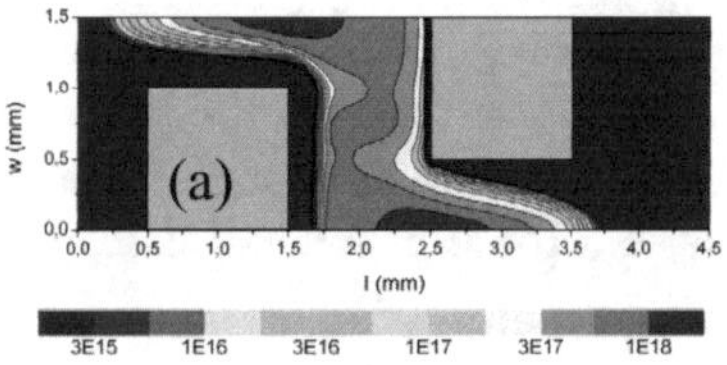

Figure 5. Electron density distribution in m^{-3} for a distance between the electrodes: 1 mm – (a), 2.5 mm – (b) and 3 mm – (c), d = 1 mm, $w_h/2$ = 0.5 mm, $w_e/2$ = 1 mm.

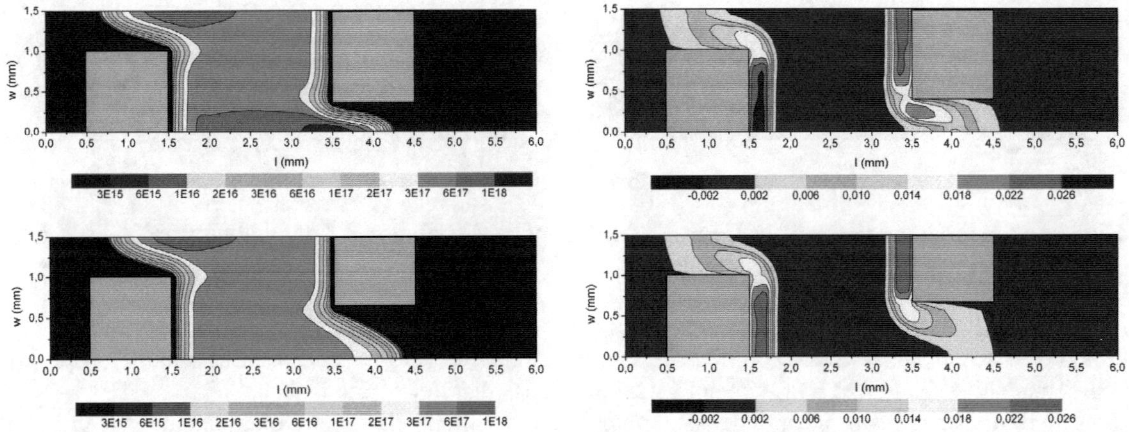

Figure 6. Distribution of electron density in m^{-3} (left) and volume charge in nC (right) for a gap spacing of 2 mm and different ratios w_h/w_e of the hole/electrode width on the right side: top 0.75/2.25 and bottom 1.25/1.75 .

1.25/1.75 are applied. The variation of the AR results in changes of discharge species density and charge distribution within the sheath regions. The plasma density (mainly within the structures with higher density) increases as the width of the holes in the right electrode decreases. The positive charge is concentrated within these holes and in the sheath region of the electrode against them (the left one). Wider holes result in lower concentration of positive charge within the holes and within the sheath of the corresponding electrode.

4.2. The effect of gas flow

In order to study the role of the gas flow in the atmospheric pressure RF discharge, calculations of the velocity field distribution have been done. The calculations are performed for unit cell with the same dimensions as mentioned in Sec. 4.1. The fluid viscosity of argon at about 350 K is computed to be $\mu = 3.5\times10^{-5}$ Pa. For calculating the initial velocities corresponding to different gas flow vessel radius r = 1.75 cm is used.

The profile obtained for inlet gas flows of 30 l/min is shown as representative of the velocity field distribution in figure 7. The velocity field has its highest absolute values at the entrances of the holes especially at the edges, within the holes and in the plasma bulk close to the grounded electrode. A backward flow toward the powered electrode is also formed in the bulk.

The effect of the flow on particle distributions is shown in figure 8. The electron density profiles shown there correspond to gas supply 30 and 60 l/min. The gas flow modifies the distributions. It leads to a reduction of the electron density at the inlet holes and an increase of the density at the outlet. For comparison see also figure 4(c). It also increases the concentration of reactive species at some distance behind the discharge.

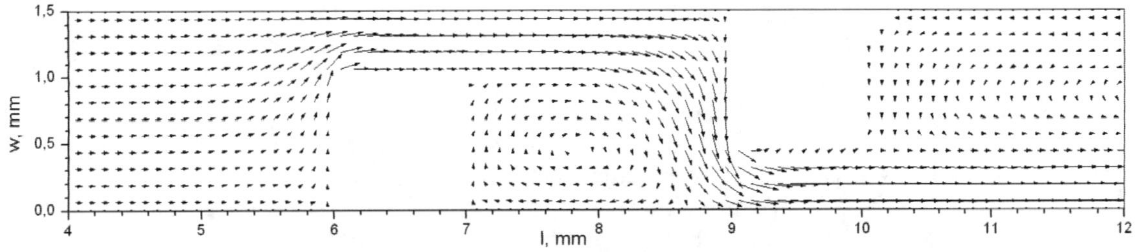

Figure 7. Spatial profiles of velocity field distribution at gas flow supply: 30 l/min.

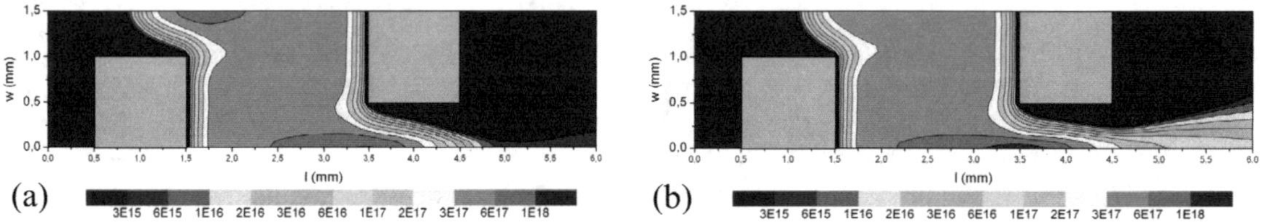

Figure 8. Spatial density profiles in m^{-3} of electrons at different gas flow velocity: (a) – 30 l/min, (b) – 60 l/min.

5. Conclusion

A model of an RF capacitively coupled discharge has been developed and used to obtain key plasma parameters and to gain insight into the physical processes that determine its behavior.

The spatial distributions of plasma density, potential and volume charge are obtained and the effect of the geometry on these characteristics is investigated.

The role of the gas flow is studied and it is found that it changes the distributions of the plasma particles between the electrodes and thus affects the plasma chemistry kinetics. The flow is found to be an effective transport mechanism to bring radicals downstream to the region where the plasma-surface interaction takes place.

6. Acknowledgements

This work was supported by Belgian Federal Science Policy programme 44 IAP-VI P6/8.

References

[1] Ji Y Y, Chang H K, Hong Y C, and Lee S H, 2009 Water-repellent improvement of polyester fiber via radio frequency plasma treatment with argon hexamethyldisiloxane (HMDSO) at atmospheric pressure, Curr. Appl Phys. 9 253.

[2] Foest R, Kindel E, Ohl A, Stieber M, and Weltmann K D 2005 Non-thermal atmospheric pressure discharges for surface modification, Plasma Phys. Control. Fusion 47 B525.

[3] Alexandrov S E, and Hitchman M L 2005 Chemical vapor deposition enhanced by atmospheric pressure non-thermal non-equilibrium plasmas Chem. Vap. Deposition. 11 457.

[4] Li H J, Wang S G, Zhao L L, and Ye T C 2004 Study on an atmospheric pressure plasma jet and its application in etching photo-resistant materials Plasma Sources Sci.Technol. 6 2481.

[5] Chandrashekar A, Lee J S, Lee G S, Goeckner M J, and Overzet L J 2006 Gas-phase and sample characterizations of multiwall carbon nanotube growth using an atmospheric pressure plasma J. Vac. Sci. Technol. A 24 1812.

[6] Li S Z, and Lim J P 2008 Comparison of sterilizing effect of nonequilibrium atmospheric-pressure He/O2 and Ar/O2 plasma jets Plasma Sources Sci. Technol. 10 61.

[7] Park J, Henins I, Herrmann H W, and Selwyn G S Selwyn 2001 Gas breakdown in an atmospheric pressure radio-frequency capacitive plasma source J. Appl. Phys. 89 1 15.

[8] Smith H B, Charles C, and Boswell R W 2003 Breakdown behavior in radio-frequency argon discharges Phys. Plasmas 10 3 875.

[9] Brok W J M, van Dijk J, Bowden M D, van derMullen J J A M and Kroesen G M W 2003 A model study of propagation of the first ionization wave during breakdown in a straight tube containing argon J. Phys. D: Appl. Phys. 36 1967.

[10] Balcon N, Hagelaar G J M, and Boeuf J P 2008 Numerical Model of an Argon Atmospheric Pressure RF Discharge IEEE Transactions on Plasma Science 36 5 2782.

[11] Shi JJ and Kong M G 2007 Mode transition in radio-frequency atmospheric argon discharges with and without dielectric barriers Appl. Phys. Lett. 90 101502.

[12] Moravej M, Yang X, Barankin M, Penelon J, Babayan S E and Hicks R F 2006 Properties of atmospheric pressure radio-frequency argon and nitrogen plasma Plasma Souces Sci. Thechnol. 15 204

[13] Moravej M, Yang X, Barankin M, Penelon J, Babayan S E and Hicks R F 2006 A radio-frequency nonequilibrium atmospheric pressure plasma operating with argon and oxygen J. Appl. Phys. 99 093305.

[14] van Dijk J, Peerenboom K, Jimenez M, Mihailova D and van der Mullen J 2009 The plasma modelling toolkit Plasimo J. Phys D: Appl. Phys. 42 194012.

[15] Hagelaar G J M, and Pitchford L C 2005 Solving the Boltzmann equation to obtain electron transport coefficients and rate coefficients for fluid models Plasma Sources Sci. Technol. 14 722.

[16] Phelps A and Molnar J 1953 Phys.Rev. 89 1202.

[17] Ellis H W, Pai R Y, McDaniel E W, Mason E A and Viehland L A 1976 Transport properies of gaseous ions over a wide energy range Atomic Data & Nucl Data Tab. 17 177.

[18] Hayashi M 1983 Determination of Electron-Xenon Total Excitation Cross Section, from Threshold to 100eV, from Experimental values of Townsend's □ J. Phys. D: Appl. Phys. 16 581.

[19] Hayashi M 2003 Bibliography of Electron and Photon Cross Sections with Atoms and Molecules Published in 20th Century: Argon, NIFS-DATA-72, National Institute for Fusion Science (Japan) ISSN 0915-6364.

[20] Rapp D, Englander-Golden P 1965 Total Cross Section for Ionization and Attachement in Gases by Electron Impact J. Chem. Phys. 43 5 1464.

[21] Vriens L and Smeets A H M 1980 Cross sections and rate formula for electron-impact ionization, excitation, deexcitation, and total depopulation of excited atoms", Phys. Rev. A 22 3 940.

[22] Balcon N, Hagelaar G J M, and Boeuf J P 2008 Numerical Model of an Argon Atmospheric Pressure RF Discharge IEEE Transactions on Plasma Science 36 5 2782.

[23] Okadat T and Sugawarat M 1993 Microwave determination of the coefficient of dissociative recombination of Ar2+ in Ar afterglow J. Phys. D Appl. Phys. 26 1680-1686.

[24] Schneider G E, Zedan M 1981 Numer. Heat Transfer 4 1–19.

[25] Hagelaar G. 2000 Modeling of Microdischarges for Display Technology Ph.D. Dissertation, Eindhoven University of Technology, Eindhoven, The Netherlands.

Journal of Physics: Conference Series **275** (2011) 012013 doi:10.1088/1742-6596/275/1/012013

Post-discharge treatment of air effluents polluted by butyl-mercaptan: role of nitrate radical

L Braci[1], S Ognier[1], YN Liu[2] and S Cavadias[1]
[1]UPMC Univ Paris 06, EA 3492, Laboratoire de Génie des Procédés et Traitements de Surface, F-75005, Paris, France
[2]School of Environmental Science and Engineering, Dong Hua University, C-201620, Shanghai, China

E-mail: Stephanie-ognier@chimie-paristech.fr

Abstract. Dry air polluted by butyl-mercaptan was treated in a Dielectric Barrier Discharge (DBD) reactor at atmospheric pressure using air as plasmagene gas in discharge and post-discharge modes. The energy density was varied between 200 to 1300 J/L. To assess the treatment efficiency, the concentrations of buty-mercaptan, total Volatile Organic Compounds (VOCs) and SO_2 were determined in the exhaust gas. Whatever the energy density was, the treatment efficiency was better in post-discharge mode. The butyl-mercaptan could be completely eliminated from 400 J/L and SO_2 selectivity was always low, below 10%. Measurements of CO, CO_2 and total VOCs indicated that 50 to 70% of the reaction products were condensed on the reactor wall in the form of micro-droplets, depending on the energy density. FTIR and XPS techniques were used to characterize the reaction products which were soluble in water. These analyses indicated that the reaction products contain oxygen, nitrogen and sulphur in an oxidized form. A reaction mechanism involving hydrogen abstraction from the -SH bond by the nitrate radical was proposed, pointing out the important role of nitrate radicals NO_3 in the reactivity of air flowing post-discharge.

1. Introduction

The mercaptans are sulfur-containing organic compounds with the general formula R-SH. They belong to the class of chemical pollutants called "Volatile Organic Compounds" (VOCs) if their vapour pressure is higher than 0.1 mmHg. These compounds are toxic and malodorous, they also represent danger for human health and ecosystem. It is thus of great importance to fight against this environmental pollution. The conventional oxidation techniques such as thermal or catalytic oxidation are not satisfactory because they require a lot of energy considering the very high conditions of temperature [1]. In the case of catalytic oxidation, the poisoning of catalyst by mercaptans is an additional problem [2]. It is therefore necessary to develop more effective technologies for the elimination of mercaptans in gaseous effluents.

Technologies based on the use of non-thermal plasma generated at atmospheric pressure are considered as relevant alternatives for the treatment of VOCs [3]. The advantage of plasma consists in its ability to generate high-energy electrons while keeping the background gas close to room temperature. Thus, a highly reactive environment is created without spending energy on gas heating as in thermal processes. Moreover, plasma techniques are flexible and easy to be implemented. Nowadays, most studies focus on processes combining plasma and heterogeneous catalysis with the objective to oxidise completely the pollutants into CO_2 with an acceptable energetic cost [4, 5]. However, such processes may not be adapted with mercaptans because of catalyst poisoning, which is the most crucial problem.

Consequently, our research team has developed an original process which has already been tested successfully with beta-pinene as a model pollutant [6]. In this process, the active species of the plasma are used to modify chemically the pollutants by partial oxidation reactions. As a result, the molecules become more polar and less volatile and in a second step, they can be trapped easily using techniques such as condensation, absorption or adsorption. With this strategy, the energy required is far lower that the energy needed for complete oxidation.

In this article the results obtained when air polluted by mercaptans is treated in discharge and post-discharge modes with the active species produced by non-thermal plasma generated at room temperature and atmospheric pressure with a Dielectric Barrier Discharge (DBD) are presented. The model sulfur molecule is butyl-mercaptan (C_4H_9-SH). To determine the treatment efficiency, the butyl-mercaptan concentration and the total VOC concentration in the exhaust gas phase have been measured. Advanced analysis techniques such as gas chromatography coupled to mass spectrometry (GC/MS) and X-ray Photoelectron Spectroscopy (XPS) have been also carried out to characterize the oxidation products. Special attention was dedicated to the comparison of discharge and post-discharge modes in order to (i) validate the best configuration and (ii) bring useful information on the species involved in the reactions.

2. Material and methods

2.1. Experimental set-up

Plasma treatment is operated in the DBD reactor as presented in figure 1. The reactor is constituted by a 190 mm long and 2 mm thick glass tube with an internal diameter of 29 mm. The stainless steel high voltage electrode is a 4 mm thick and 25 mm diameter disc. The space between high voltage electrode and the glass tube used as dielectric material is 2 mm. A wire sheet (20 mm width and 2 mm thick) wrapped around the tube constitutes the ground electrode. A high voltage generator supplies from 5 to 30kV peak to peak (U_{pkpk}) alternative voltage. The voltage and the current are measured by a LeCroy LT 342, 500 MHz digital oscilloscope. In the conditions of the experiments, the discharge is filamentary. The discharge power P was calculated by the integration on two entire periods of instant applied voltage and instant discharge current. The energy density DE (J/L) was defined as the ratio P/Q where Q is the gas flow (L/s) through the discharge. During the experiments, the discharge power was varied from 2 to 24 watts (DE from 100 to 1200 J/L).

1.2 L/min of pure air is injected at the gas inlet A to flow between the two electrodes. The synthetic polluted gaseous effluent containing 36 000 ppmv of butyl-mercaptan is produced by bubbling 40 ml/min of air in liquid butyl-mercaptan. This polluted air flow is either premixed with pure air and injected at the gas inlet A (“discharge” mode) or injected after the discharge zone at the gas inlet B (“post-discharge” mode). The distance of the gas inlet B from the electrode is 10 mm. The air flow carrying the pollutant is low compared to the pure air flow so that the gas flow through the discharge could be considered identical whatever the location of pollutant injection is.

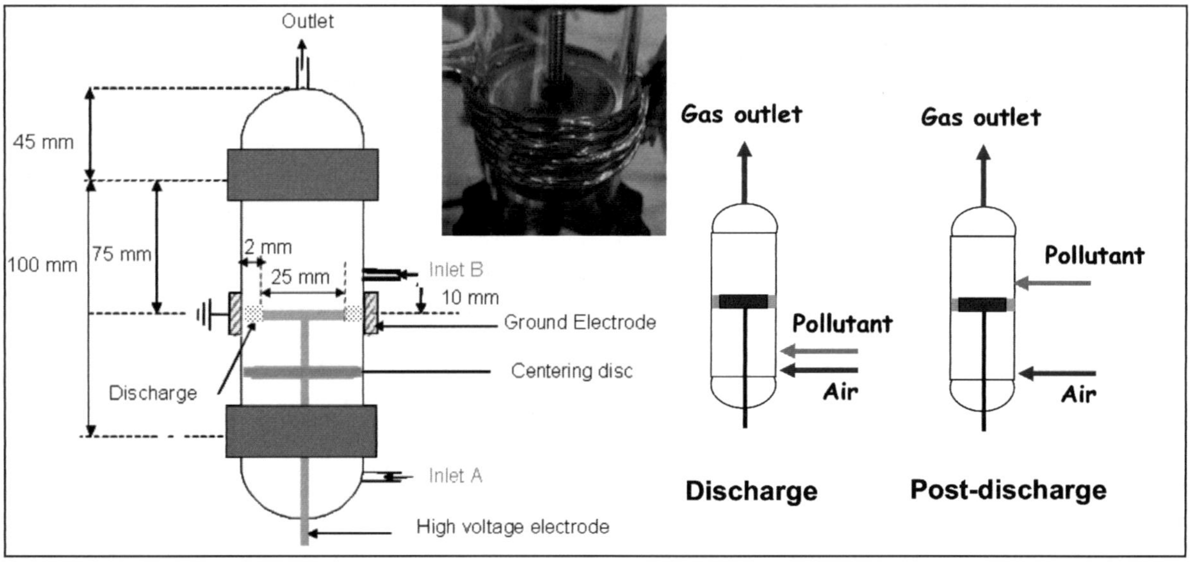

Figure 1. DBD reactor and configurations tested

2.2. Analytical methods

Two types of analyses were carried out during the study: the analyses of the exhaust gas phase and the analyses of the products condensed on the reactor wall.

2.2.1. Analyses of exhaust gas

Gaseous samples were taken at the exit of the plasma reactor and injected in a gas chromatograph (Varian STAR, 3400Cx) equipped by a Flame Ionization Detector (FID). The injector and detector temperatures were set at 250 °C and high-purity He gas was used as the carrier gas. To determine the pollutant concentration, a capillary non polar column (BPX5, SGE) was used. To measure the concentration in total Volatile Organic Compounds (total VOCs), a short silica capillary (length = 10 cm) heated at 150°C and connected directly from the injector to the detector was used. In this case, the organic compounds are burned together in the flame (no separation) and one peak corresponding to the sum of the atoms of organic carbon present in the sample is obtained.

The concentration of Sulphur Dioxyde (SO_2) was determined according to the protocol of standard NF X43-013. In this method, the SO_2 contained in the exhaust gas is absorbed in a solution containing H_2O_2 and dioxane to be converted into sulphate ions SO_4^{2-}. The sulphate ions are then measured by a spectrophotometric method using thorin.

2.2.1. Characterization of oxidation products condensed on the reactor wall

During the experiment, the formation on the reactor wall of a white deposit consisting of micro droplets was observed. For FTIR analysis, the deposit was recovered in CH_2Cl_2 and directly analysed. X-ray Photoelectron Spectroscopy (XPS) measurements were carried out to characterize the chemical composition of the oxidation products and the electronic character of sulphur element. In this method, a thin film of pure Polyethylene (PE) is pasted on the upper part of the reactor interior wall. The plasma treatment is then operated during 15 minutes, allowing the surface of the PE film to be covered by the treatment products. Finally, the PE film is removed and XPS analysis of its surface is carried out (XPS PHI 5600-ci, Physical Electronics, Eden Prairie, MN, USA) with Al Ka (1486.6 eV) source for spectrums on survey scan (chemical compositions) and Mg Ka (1253,6 eV) source for high resolution spectrums. The spectrometer was run between 0 and 1400 eV during 8 min for survey spectrums. High resolution spectrums on carbon (C) and sulfur (S) were also conducted using 10 and 20 scans respectively. A correction for binding energy was made to account for sample charging based on the C1s peak at 285.0 eV. The S2p peak curve was resolved using mixed Lorentzian-Gaussian line shape.

3. Experimental results

3.1. Air treatment efficiency

3.1.1. Butyl-mercaptan and VOCs removal efficiencies

The concentrations of butyl-mercaptan in the exhaust gas were measured for treatments carried out in discharge and post-discharge modes at different energy densities. The figure 3 shows the butyl-mercaptan removal as function of the applied energy density for the two configurations.

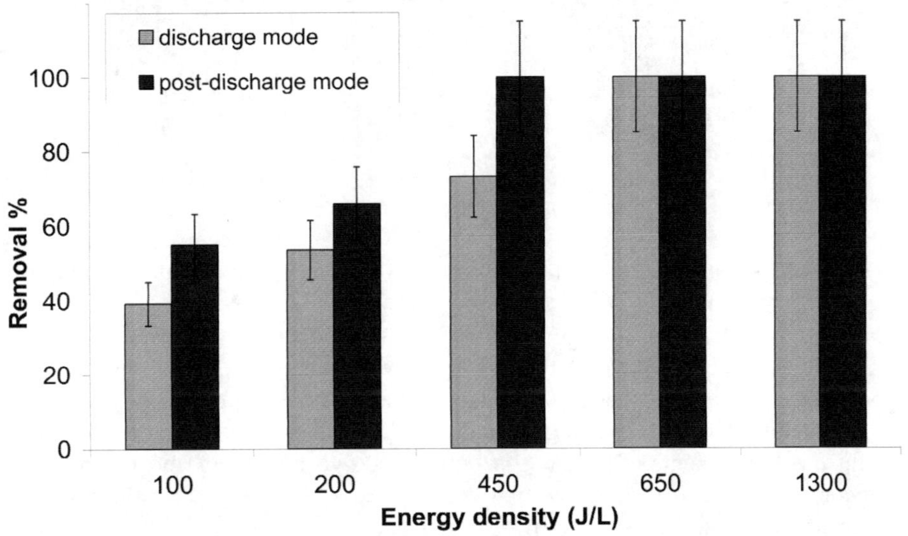

Figure 3. Butyl-mercaptan removal vs energy density in discharge and post-discharge

The butyl-mercaptan removal increases logically when the energy injected per liter of gas increases, indicating that the number of active species which can react with butyl-mercaptan increase when the energy density increases. Surprisingly, for a given energy density, the removal is slightly higher in the post-discharge mode compared to the discharge mode. For example, at 450 J/L, the removal is 100% in post-discharge mode but only 73 % in discharge mode. In post-discharge mode, only the long-life active species produced by the discharge can react with the target pollutant whereas in discharge mode, the short-life species such as atomic oxygen can also participate to the oxidation reactions. The respective roles of short-life and long-life active species will be discussed in section 4.

The total VOCs concentrations in the gas exiting the reactor were also determined for different energy densities. Useful information concerning the fate of the reaction products can be obtained when measuring total VOCs concentration. In our experimental conditions, no CO or CO_2 was measured indicating that the oxidation of the pollutant was not complete. In this case, if the oxidation products are volatile enough to remain in the gaseous state, the total VOCs concentration should remain unchanged. On the contrary, if the volatility of the oxidation products is low enough so that they can condense on the reactor wall, the total VOCs concentration should decrease. The figure 4 presents the removal of total VOCs defined as $\frac{[\text{total VOCs}]^{\text{discharge off}} - [\text{total VOCs}]^{\text{discharge on}}}{[\text{total VOCs}]^{\text{discharge off}}}$ as function of the energy density.

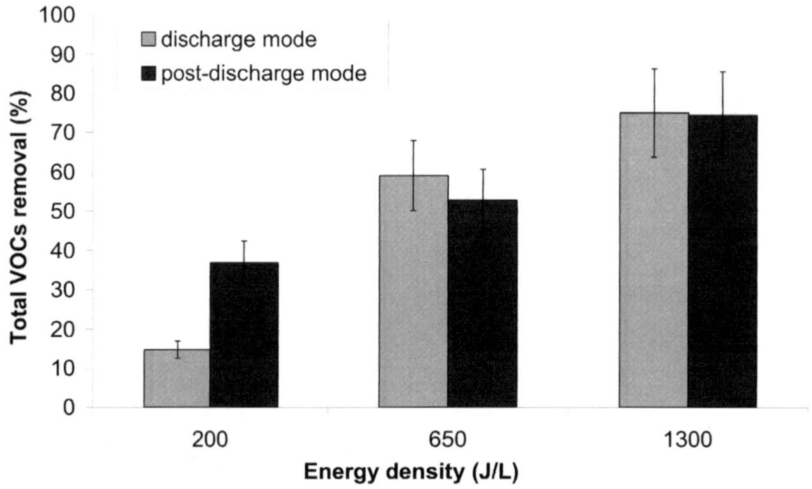

Figure 4. Relative concentrations of total VOCs as function of energy density

The removals of total VOCs are not equal to zero, which indicates that some of the carbonaceous products formed by the reaction of the butyl-mercaptan with the active species are no longer gaseous. These compounds are probably responsible for the formation of droplets on the reactor walls. In fact, the formation of a white deposit consisting of micro droplets in the upper part of the reactor wall was observed during the experiments. The reaction between butyl-mercaptan and plasma species is then responsible of the formation of products which are spontaneously trapped on the wall of the reactor. With these measurements, the efficiency of trapping was calculated. The trapping efficiency is the amount of organic carbon contained in the reaction products condensed on the reactor walls divided by the total amount of organic carbon contained in all the reaction products. It can be calculated by the following expression:

$$\textit{Trapping efficiency} = \frac{[\text{total VOCs}]^{\text{discharge off}} - [\text{total VOCs}]^{\text{discharge on}}}{[\text{total VOCs}]^{\text{discharge off}} \times (1 - \tau_{\text{butanethiol}})}$$

Where $\tau_{\text{butanethiol}}$ is the butyl-mercaptan removal. The variation of the trapping efficiency as function of the energy density is represented in table 1.

Table 1. Trapping efficiency

	200 J/L	*650 J/L*	*1300 J/L*
Discharge mode	15	57	75
Post-discharge mode	37	53	74

The trapping efficiency is about 50% for low (200 J/L) and medium (650 J/L) energy densities. For 1300 J/L, a higher value of 70% is obtained. At high energy density, the higher amount of active species compared to the amount of pollutant may allow the formation of more condensable products.

3.1.2. SO_2 formation

The sulfur dioxide SO_2 was measured in the exhaust gas and the SO_2 selectivity was calculated. It is defined as the ratio between the number of moles of SO_2 that have been formed n_{SO_2} and the number of moles of butyl-mercaptan that have reacted $n_{\text{butyl-mercaptan}}$:

$$SO_2\ selectivity = \frac{n_{SO_2}}{n_{\text{butyl-mercaptan}}} = \frac{[SO_2]^{out}}{[Butyl-mercaptan]^{in} - [butyl-mercaptan]^{out}}$$

Where $[SO_2]^{out}$ is the molar concentration of SO_2 at the reactor outlet, and $[\text{butyl-mercaptan}]^{in}$ and $[\text{butyl-mercaptan}]^{out}$ the molar concentrations of butyl-mercaptan at the inlet and at the outlet of the reactor. Table 2 shows the SO_2 selectivity for each energy density applied.

Table 2. SO_2 selectivity as function of energy density

	200 J/L	*650 J/L*	*1300 J/L*
Discharge mode	9	11	37
Post-discharge mode	4	4	10

Whatever the energy density, the SO_2 selectivity is low (< 10%) what shows that the butyl-mercaptan is only partially oxidised. The SO_2 selectivity remains almost constant between 200 and 650 J/L with a value of about 3% then increases abruptly to reach 9% at 1300 J/L.

3.2. Characterization of the oxidation products condensed on the reactor wall

Two different techniques were used to characterize the oxidation products trapped on the reactor wall. Whatever the technique used, there was no significant differences between the two configurations. Consequently, only the results obtained in post-discharge mode are presented.

3.2.1. FTIR analysis

FTIR analyses were carried out to obtain structural information about the oxidation products trapped on the reactor wall. On the spectrum presented in figure 5, the presence of C-H bonds is shown by the three bands at 2874, 2930 and 2960 cm^{-1} and the band at 1466 cm^{-1}. The broad absorption band around 3300 cm-1 may correspond to the OH function of a carboxylic acid. At 1735 cm-1, the characteristic band of C=O function is detected. In the presence of ester compounds, a strong C-O absorption band is detected around 1250-1150 cm^{-1}. The bands around 1200 cm^{-1} may indicate that the deposit contains ester compounds. On the spectrum, a strong band is visible at 1040 cm^{-1}. This band may be attributed to the C-O bond in a carboxylic acid or an ester, but in this case its intensity is low compared to the two other vibrations observed at 1735 and 1250-1150 cm^{-1}. We can therefore assume that another chemical group is at the origin of this band. This band could be due to oxygenated sulfur compounds such as sulfoxides RS (= O)-R '(strong absorption between 1030 and 1070 cm^{-1}) or sulfonate ion (1055 and 1175 cm^{-1}). Some bands characteristic of other oxygenated sulfur compounds are also visible. The bands at 1130 and 1320 cm^{-1} may be due to sulfones $RS(=O)_2$-R'. Similarly, the bands at 1180 and 1380 cm^{-1} can be attributed to compounds of type RS(= $O)_2$-OR'. Finally, the two intense bands at 1630 and 1280 cm^{-1} correspond to the vibration of the NO_2 group in an organic nitrate RO-NO_2.

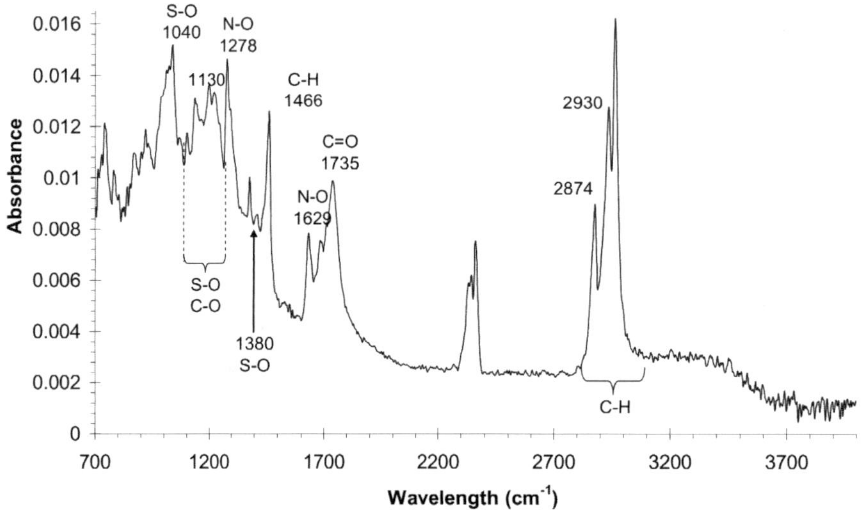

Figure 5. FTIR spectrum

3.2.2. XPS analysis

A survey scan was first carried out to obtain the elemental composition of the surface of the polyethylene film covered by the oxidation products. The composition was as follows: carbon 83.3%, oxygen 11.2%, nitrogen 3.6% and sulfur 3%. These values do not represent the average chemical composition of the oxidation products because the PE was only partially covered. However, these results confirm the presence of nitrogen and sulfur in the oxidation products. In a second step, high resolution XPS measurements were carried out to obtain more information on the chemical bonding of the sulphur element in the oxidation products. The deconvolution of the XPS spectrum of sulphur S_{2p} presented in figure 6 gives two chemical states of sulphur with two peaks at 169,84 eV and 168,54 eV. The full width at half maximum (FWHM) was fixed at 1.95 eV [7]. These binding energies correspond to a high degree of oxidation of sulphur. The functions proposed for a binding energy greater than 168 eV are in increasing order of binding energy: R-SO_3^-, R-SO_2-SR', R-SO_2-OR', SO_4^{2-},-SO_3-ROO and R-SO_2-OH [8].

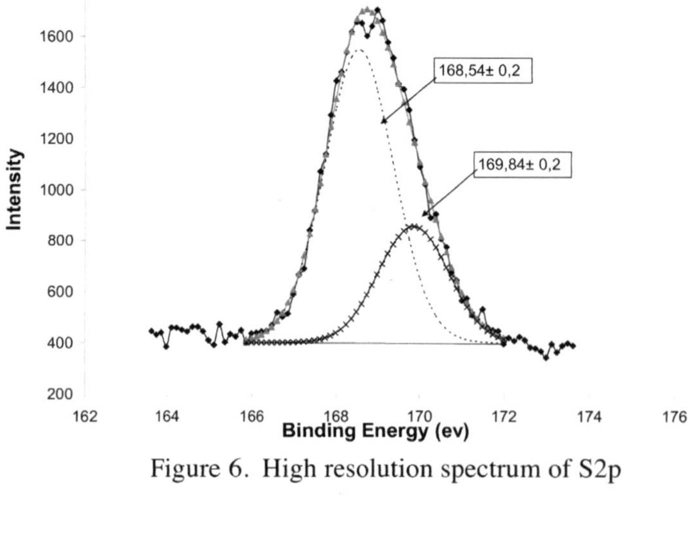

Figure 6. High resolution spectrum of S2p

4. Discussion

The experimental results indicate that the long-life species play a major role in the oxidation of the butyl-mercaptan. In fact, the results obtained in the discharge and post-discharge modes are roughly the same in terms of butyl-mercaptan removal, total VOCs removal, SO_2 formation, and nature of oxidation products.

The long-life species produced by an atmospheric plasma discharge in air have been characterized by Eliasson and Kogelschatz [9]. A single microdischarge in air with a short current pulse of 10 ns duration deposits energy in various excited levels of N_2 and O_2, some of which lead to dissociation and finally to the formation of ozone and different nitrogen oxides. After about 50 ns, most charge carriers have disappeared and after 10 ms, only long-life species such as O_3, N_2O, NO, NO_2, NO_3 and N_2O_5 remain. At low energy density, the concentrations of NO_2, NO_3 and N_2O_5 are low compared to the concentration of O_3. On the opposite, at high energy density, the concentration of O_3 become negligible compared to the concentrations of nitrogen species.
Information about reactivity of the long-life species with mercaptans can be found in the literature dealing with atmospheric chemistry [10]. Among the long-life species produced by the atmospheric air discharge, only the nitrate radical NO_3 can react with mercaptans. The kinetic constants of the second-order reactions between NO_3 and ethyl-mercaptan C_2H_5-SH and methyl-mercaptan CH_3-SH are respectively $1.2E^{-12}$ [10] and $9.2E^{-13}$ [11] $cm^3.molecule^{-1}.s^{-1}$ at 298 K and the reaction mechanism involves hydrogen abstraction from the -SH bond by the nitrate radical with formation of a radical R-S° and nitric acid HNO_3 [12]:

$$R\text{-}SH + NO_3 \rightarrow R\text{-}S° + HNO_3 \quad (1)$$

The radical R-S° can then undergo multiple reactions due to reactions with molecular oxygen and/or nitrogen oxides. Basing on the mechanism proposed by Jensen et al.[12], the scheme presented in figure 7 can be suggested for the destruction of mercaptans:

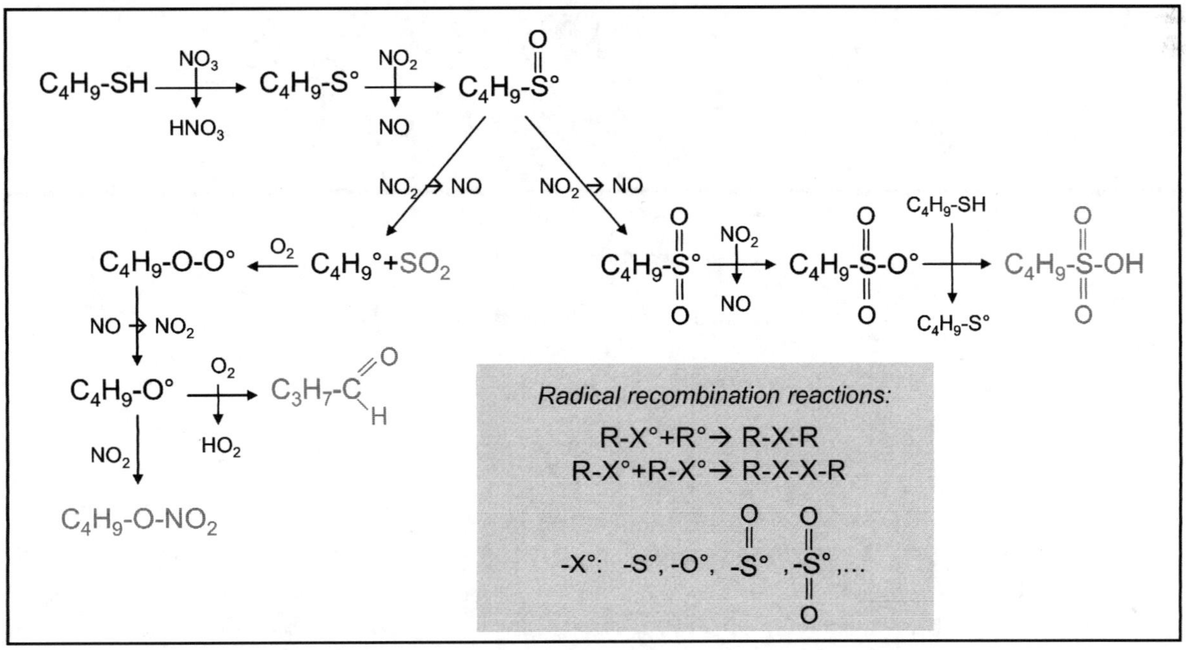

Figure 7. Possible mechanisms for the oxidation of C_4H_9SH initiated by NO_3

According to this reaction scheme, the final reaction products are butanesulfonic acid R-S(=O)$_2$-OH, aldehydes, butyl nitrate R-O-NO_2 and SO_2. In the conditions of the present study, the concentration of butyl-mercaptan is high and recombination reactions between radicals (RS°, RSO°, RSO_2°...) are also likely to occur with production of high molecular weight compounds.
This mechanism is coherent with the FTIR and XPS analyses presented in 3.2. FTIR analyses indicated the presence of organic compounds with S-O or S=O functions, aldehyde and/or ester functions as well as -O-NO_2 groups and XPS analyses highlighted the high content of sulphur and nitrogen in the reaction products as well as the high oxidation state of sulphur.
It can be noted that sulfonic acids and high molecular weight compounds are likely to condense easily either in a homogeneous way or in a heterogeneous way. This explains the condensation as micro-droplets on the reactor wall [13].

The crucial role of long-life species brings also fundamental information on the phenomena taking place in discharge mode. In air at atmospheric pressure, the electrical discharge is filamentary. In the gaseous gap between the two electrodes, we can distinguish two regions [14]: the filamentary volumes where discharge takes place ("discharge" region) and the region outside these volumes ("post-discharge" region) (figure 8). Each time the discharge takes place, electrons and active species such as ions, radicals and excited species are generated in the filamentary volumes. They recombine rapidly into more stable species and can further diffuse in the post-discharge region. The short life duration of radicals such as atomic oxygen O° makes that in the post-discharge region, only the molecules in the close vicinity of the filaments are likely to react with them. As a consequence, if the ratio (filamentary volume/total volume) is low, the reactions involve mainly the long-life species. In the conditions of the present study, it can be therefore supposed that the (filamentary volume)/(total volume) ratio is low.

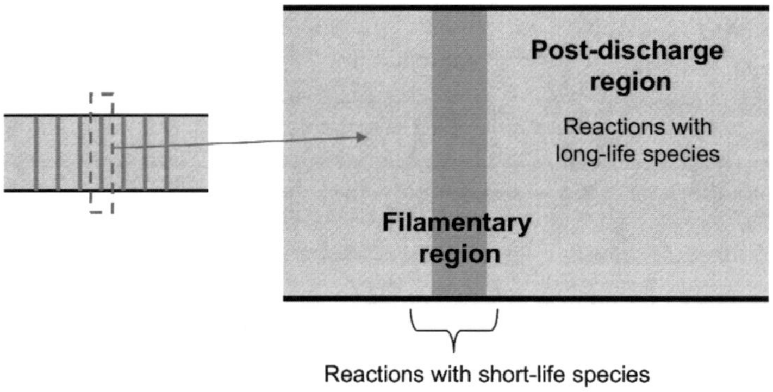

Figure 8. Schematic of the idealized discharge structure

The experimental results have also shown that for a given energy density, the removal of butyl-mercaptan is slightly higher in the post-discharge mode compared to the discharge mode. To explain this result, we can introduce the yield of the oxidation of butyl-mercaptan defined as follows:

$$Yield = \frac{R - SH \text{ removed by the discharge}}{O° \text{ formed by the discharge}}$$

In the post-discharge mode, the oxygen radicals in the filamentary region are converted partly in NO, NO_2 and NO_3. The NO_3 radicals can react then further in post-discharge region with butyl-mercaptan according to (1). This reaction is selective because the rate constants of NO_3 radicals toward oxidation products such as aldehydes, sulfonic acids or R-O-NO_2 are low compared to the rate constant toward butyl-mercaptan. As a result, a high yield is obtained.

Journal of Physics: Conference Series **275** (2011) 012013 doi:10.1088/1742-6596/275/1/012013

In the discharge mode, some oxygen radicals are converted in NO and NO_2 and NO_3 as previously mentioned, whereas other ones react directly not only with butyl-mercaptan but also with its oxidation products. In fact, O° radicals are nonselective toward organic species. As a consequence, in discharge mode, less NO_3 is produced and butyl-mercaptan is oxidised with a low yield in the filamentary region, which explains why butyl-mercaptan removal is slightly higher in the post-discharge mode compared to the discharge mode.

5. Conclusion

In this work, air polluted with butyl-mercaptan was treated with non-thermal DBD plasma in discharge and post-discharge modes. In the two modes, the oxidation reactions took place mainly in post-discharge region and the nitrate radicals played a crucial role in the oxidation of butyl-mercaptan. Spontaneous condensation of the oxidation products on the reactor wall was observed, which indicates that the plasma treatment could be used as a pre-treatment to modify the properties of the mercaptans so that they can be trapped further easily using techniques such as condensation, absorption or adsorption. Finally, the results show that if the plasma treatment is carried out with the objective to oxidise partially the mercaptans, the post-discharge mode is the best configuration.

References

[1] Li WB, Wang JX, Gong H 2009 Catal. Today 148 1-2 81-87
[2] H, Hao GH, Tseng TK 2003 J. Hazard. Mater. 100 1-3 301-16
[3] Parmar GR, Rao NN 2009 Crit. Rev. Environ. Sci. Technol. 39 1 41-78
[4] Van Durme J, Dewulf J, Leys C, Van Langenhove H 2008 Appl. Catal. B:Environ. 78 324-33
[5] Chen HL, Lee HM, Chen SH, Chang MB, Yu SJ, Li SN 2009 Environ. Sci. Technol. 43 2216-27.
[6] Ognier S, Youssef J, Cavadias S, Amouroux J 2008 Int. J. Chem. React. Eng. 6 A27.
[7] Thirumaran S, Ramalingam K, Bocelli G, Righi L 2009 Polyhedron 28 2 263-268
[8] Siow KS, Britcher L, Kumar S, Griesser HJ 2009 Plasm. Process. Polym. 6 583-92
[9] Eliasson B, Kogelschatz U 1991 IEE Transactions on Plasma Science 19 309-23.
[10] Atkinson R, Baulch DL, Cox RA, Crowley JN, Hampson RF, Hynes RG, M. E. Jenkin ME, Rossi MJ, Troe J 2004 Atm. Chem. Phys. 4 1461-1738
[11] Mac Leod, H, Aschmann SM, Atkinson R, Tuazon EC, Sweetman JA, Winer AM, Pitts JN 1986 J. Geophys. Res 91
[12] Jensen NR, Hjorth J, Lohse C, Skov H, Restelli G 1992 Int. J. Chem. Kin. 24 10 839-50
[13] Vandingenen R, Jensen NR, Hjorth J, Raes F 1994 J. Atm. Chem. 18 3 211-37.
[14] Redolfi M, Aggadi N, Duten X, Touchard S, Pasquiers S, Hassouni K 2009 Plasm. Chem. Plasm. Process. 29 3 173-195.

Journal of Physics: Conference Series **275** (2011) 012014 doi:10.1088/1742-6596/275/1/012014

Development of a hollow cathode plasma source for microcrystalline silicon thin films deposition

P Dimitrakellis, E Amanatides, D Mataras[1] and DE Rapakoulias
Plasma Technology Laboratory, Department of Chemical Engineering, University of Patras, P.O. Box 1407, 26504 Patras, Greece

E-mail: dim@plasmatech.gr

Abstract. The development of a hollow cathode plasma source and its implementation in an already existing plasma reactor used for the deposition of μc-Si:H thin films is presented. Electrical measurements of hydrogen discharges at 13.56 MHz were carried for two different in geometry hollow electrodes and the electrical parameters were compared with a previous Capacitively Coupled source installed in the same reactor. The results show a significant enhancement of discharge current and higher power dissipation for the same applied voltage. At reduced gas pressures that are necessary to minimize the dust formation, the hollow electrode with the larger cavity geometry showed better performance, achieving high electron densities of above 10^9 cm^{-3}. Microcrystalline silicon thin films were deposited from SiH_4/H_2 discharges and an enhancement of the growth rate was found for the hollow cathode source as compared with the CCP source.

1. Introduction

Hydrogenated microcrystalline silicon (μc-Si:H) thin films have attracted intense scientific interest due to their application in thin film solar cells of single and tandem structure [1-3]. Although this material presents wider-spectrum response and enhanced stability against light exposure than the technologically mature hydrogenated amorphous silicon (a-Si:H), thicker films of μc-Si:H are required to sufficiently absorb the light due to its indirect band gap [1,4]. Thus, an issue that emerges inevitably is how the film deposition process can be less time-consuming, aiming to reduce the production costs, while preserving the desired film quality for application in photovoltaic industry. The most widespread method used industrially for deposition of μc-Si:H is Capacitively Coupled silane/hydrogen glow discharges (CCP-GDs), though with this technique the growth rate for device grade films is rather low. This is the main reason why the research is directed to the implementation of different, non-conventional plasma sources with enhanced electron density, which can lead to higher gas usage efficiencies and deposition rates. Many novel plasma techniques have been proposed in this field, such as the very high frequency (VHF) and the high pressure depletion method (HPD) [5-7]. Other high density methods using microwave plasma [8,9], inductively coupled plasmas [10,11] and electron cyclotron resonance plasmas [12,13] have also been proposed. Hollow cathode discharges (HCDs) present noticeably high electron density due to the electrons entrapment and oscillation inside narrow cavities and are promising for deposition processes [14-17]. The advantage of hollow cathode against other high density plasma sources is that only small changes are needed on the already existing industrial reactors. Hollow cathode sources for μc-Si:H deposition can be found in literature in diverse configurations with very interesting results concerning the deposition rate and the film quality [18,19].

Other interesting issues related to hollow electrodes can also be found, such as electrical measurements, optical emission spectroscopy analysis and quality evaluation of films deposited at high growth rates [20-22].

In this work, we present the development of a hollow cathode (HC) electrode and the implementation of the source in an already existing parallel plate Capacitively Coupled plasma reactor. The basic aims of the study are to investigate the role of hollow geometry on the deposition rate of μc-Si:H thin films as well as to study the film growth rate under relatively low total gas pressure in order to minimize the problems of dust formation. For that reason we developed two different in geometry hollow cathode sources and electrical measurements were performed for both of them, in hydrogen discharges. The measurements of voltage, current and phase impedance in the plasma and the calculation of discharge power revealed the conditions required for operation in the hollow cathode mode. A comparison of the different sources with the existing CCP source was also implemented and the differences in the power dissipation and the electron densities are presented. Finally, thin silicon films depositions were performed as a function of various experimental parameters and the changes on the film growth rates for the HCD and the CCP source are discussed.

2. Experimental

The experiments were carried in an ultrahigh vacuum reactor having a base vacuum of 10^{-7} Pa. The 12 cm in diameter RF powered hollow cathode electrode is fixed to the chamber in the place of the already existing capacitively coupled planar electrode and the measurements were performed in two different in geometry hollow electrodes, the first (HC1) consisting of 40 holes of 0.5 cm length and 3 cm diameter (Figure 1) and the second (HC2) of 12 holes of 1.8 cm length and 2 cm diameter (Figure 2). The substrate holder (grounded electrode) can be freely moved to adjust the distance between the HC source and the substrate. All the films were deposited in Corning 7059 at substrate temperature of 200 °C using high purity gases and the deposition rate was measured *in-situ* using Laser Reflectance Interferometry (LRI).

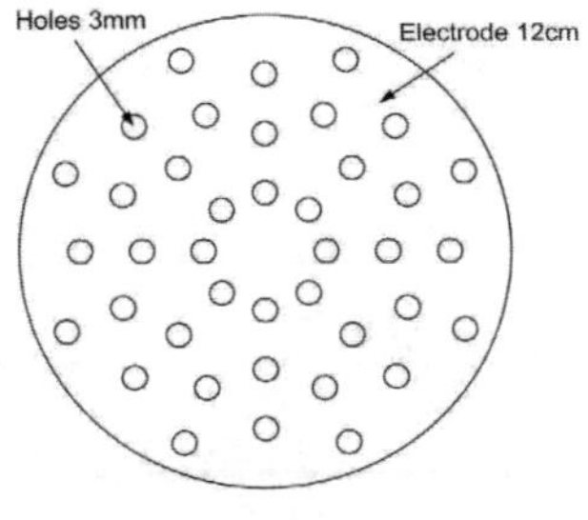

Figure 1. HC1

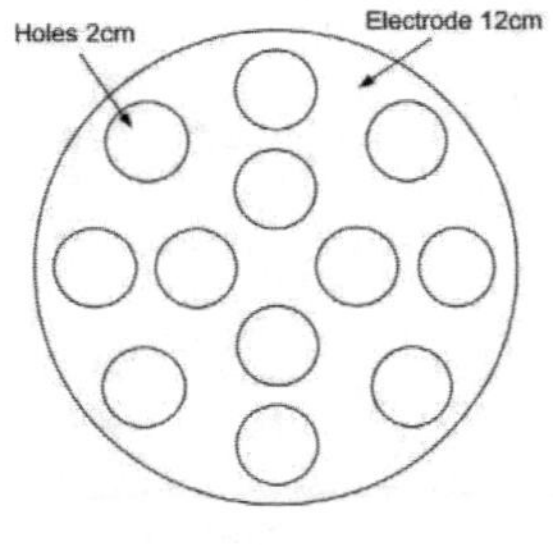

Figure 2. HC2

The RF (HC or CCP) electrode was powered by a 13.56 MHz generator, through an L-type matching network. A Lecroy PPE 1000:1 voltage probe and a FCC F-35-1 current probe were attached after the matching network. A Lecroy 9400 digital oscilloscope was used to record the voltage and current signals which were then transferred to a computer for Fourier analysis. The real amount of RF power consumed in the discharge was determined using an accurate method described in details elsewhere [23]. With this method we were able to convert the voltage and current waveforms measured outside the chamber to equivalent waveforms at the RF electrode [24].

All sets of electrical measurements were performed for pure H_2 discharges while the total gas pressure and the flow rate were kept constant at 133 Pa and 50 sccm respectively. The hydrogen was fed to the reactor through the hollow cathode electrode. The distance between the powered and the grounded electrode was set to 1.5 cm. As for the deposition experiments, the total gas flow rate was maintained constant at 200 sccm while the silane ratio was set at 10 %. The effect of pressure (79.8 to 332.5 Pa), interelectrode space (1.5 to 2.5 cm) and applied voltage (50 V to 150 V amplitude) was investigated. Partial flow rates were adjusted by accurate mass flow controllers and the total pressure was controlled by a downstream throttle valve controller as well.

3. Results and discussion

3.1 Electrical measurements

Different electrical parameters were examined on hydrogen discharges comparing the three different plasma sources and configurations; the typical capacitively coupled glow discharge with the two other, different in geometry, hollow cathode sources. The discharge current as a function of the electrode voltage amplitude is shown in Figure 3. In the CCP case the current increases slightly with the voltage with its values limited in a range of 100-200 mA. On the other hand, in the hollow cathode discharges and for both hollow electrodes the current showed a sharp, almost exponential increase with the electrode voltage. Additionally, the second hollow configuration (HC2) presented much higher current values than the two other sources, indicating conditions of higher electron density. It is also worth noticing that for HC2 the current increases abruptly after the 100 V, revealing the breakdown voltage for which the discharge moved from the early capacitive to the hollow cathode mode.

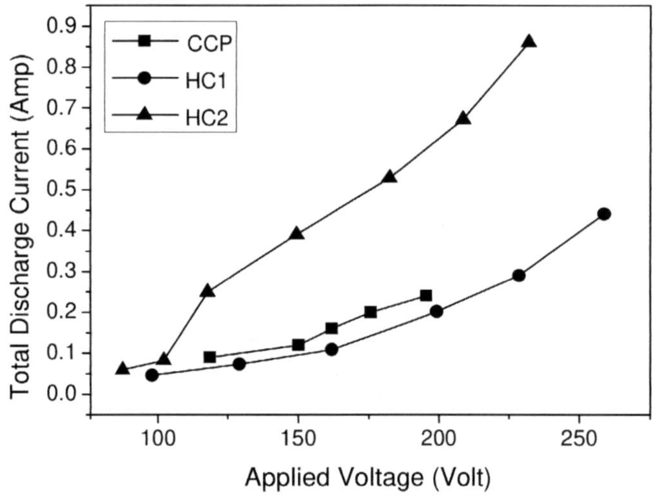

Figure 3. The total discharge current as a function of the electrode voltage amplitude for three different plasma sources.

In this mode, the increase of the current is attributed to the hollow cathode effect that leads to avalanche multiplication of charge carriers inside the cavities of the electrode. Comparing the two hollow cathodes it is clear that for HC1 there is no mode transition and operates in a purely capacitive mode. The breakdown voltage is totally affected by the hollow geometry and the pressure and follows Paschen's Law [15]. Thus, for the first electrode configuration with the narrow holes, higher pressures are prerequisite for the mode transition, while the wide cavities in the second hollow electrode allow process optimization at relatively low gas pressures.

Figure 4 presents the real power dissipation in the discharge as a function of the electrode voltage amplitude. The variation of power presents similar behaviour to the discharge current apart from the fact that the values concerning the CC and the HC1 electrode almost coincide, confirming that in those conditions the latter behaves as being in a capacitive mode. On the other hand, the sharp increase of

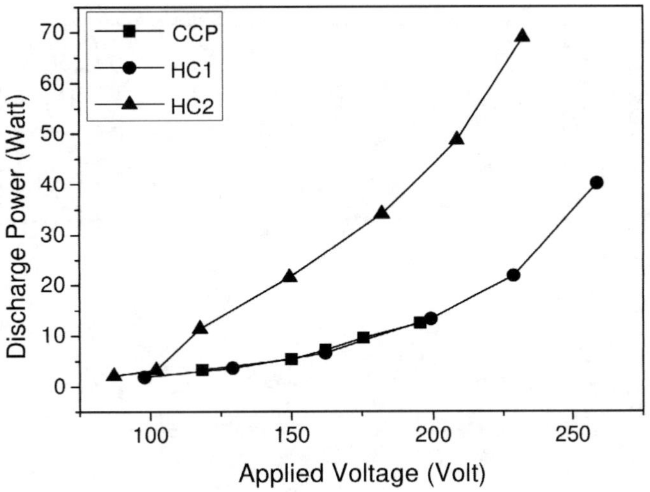

Figure 4. The total discharge power as a function of the electrode voltage amplitude for three different plasma sources.

the discharge power with voltage for HC2 follows the increase of the current that was shown in fig. 3.

The variation of applied voltage and modes affects also the reactive and ohmic part of the plasma impedance. The complex discharge impedance is given by the equation [23]:

$$Z = \frac{V_{el}}{I_{el}} \exp(i\varphi_{el}) = R + iX \quad (1)$$

where φ_{el} is the phase of the electrode voltage V_{el}, relative to the electrode current I_{el}. The ohmic component R of the impedance is given by:

$$R = Z\cos\varphi_{el} = \frac{V_{el}}{I_{el}}\cos\varphi_{el} \quad (2)$$

and counts for the power dissipation in the discharge due to the acceleration of electrons in the plasma bulk and also of ions and secondary electrons in the sheaths. Figure 5 presents the variation of R as a function of the applied voltage for the three different sources. The ohmic component drops with voltage for all modes and geometries as a result of the increase of plasma conductivity. This reduction is more pronouced in the case of the second hollow electrode where the impedance falls abruptly at the mode transition region.

The information acquired by the electrical measurements concerning macroscopic parameters can be linked to microscopic parameters in the plasma such as electron number density, energy and collision frequency. The electron density is an essential quantity when one argues about high density plasma sources. An estimation of the average electron density can be given from the electrical measurements as it is strongly related to the ohmic part of the discharge impedance R [24-26]:

$$n_{AV} = \frac{v_m d_b m_e}{RAe^2} \quad (3)$$

where v_m is the electron momentum transfer collision frequency, d_b is the bulk length, m_e is the electron mass, A is the discharge cross section and e is the electron charge. As bulk length was considered the whole interelectrode distance (1.5 cm) and the momentum transfer collision frequency was estimated at ~3.76 x 10^7 $Pa^{-1}s^{-1}$ [25]. Figure 6 presents the variation of the average electron density as a function of the electrode voltage amplitude. The comparison of HC1 with CC source shows that the electron density in the latter is slightly higher, which in turn reveals that in that pressure and electrode

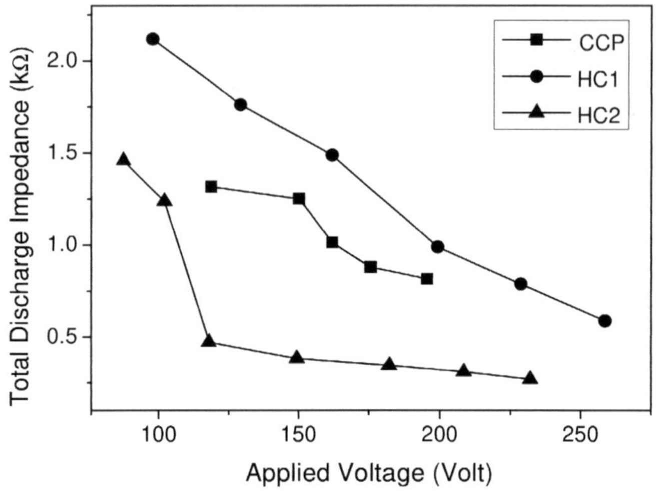

Figure 5. The discharge impedance as a function of the electrode voltage amplitude for three different plasma sources.

gap the hollow cathode electrode not only operates in a typical capacitive mode but it's even less efficient. On the other hand, the average electron density in HC2 is 3 to 4 times higher than both HC1 and CC configurations, reaching values higher than 10^9 cm^{-3}. In this mode, inside the cavities the electrons multiply rapidly beacause the sheath voltage increases leading to higher electric field. Electrons in the bulk are trapped between repelling electric fields of the opposite sheaths and are forced to oscillate, gaining more energy and causing ionization collisions with the hydrogen molecules. In addition, the higher sheath voltages accelerate the ions towards the hollow cathode walls causing secondary electron emission, which contributes to the multiplication of electrons and the increase of their number density. However, the increase of the electron density with voltage can also be indirectly affected by pressure, which is higher inside the cavity than in the reactor chamber [15].

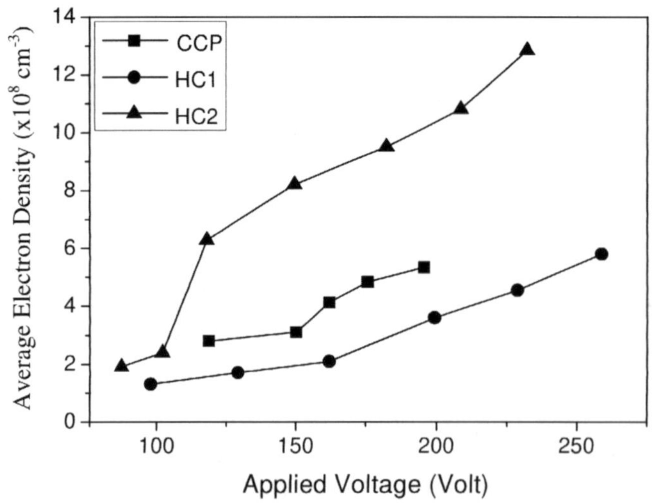

Figure 6. The average electron density as a function of the electrode voltage amplitude for three different plasma sources.

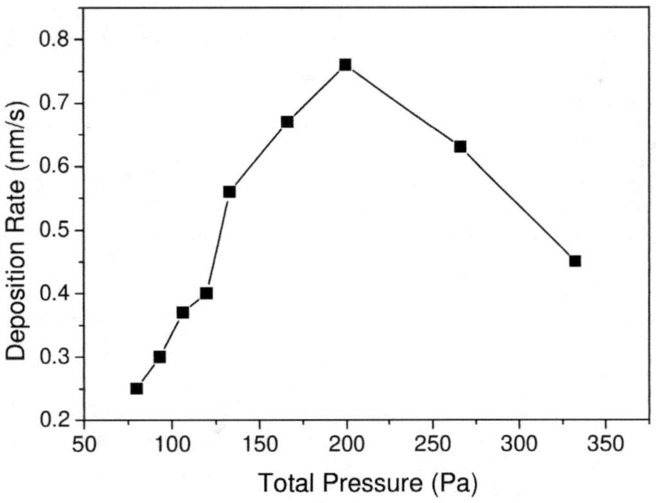

Figure 7. The deposition rate as a function of total gas pressure.

This means that a pressure variation inside the hollow cathode, which affects the ionization collision frequency, may contribute to the variation of the average electron density.

3.2 Deposition rate

Microcrystalline silicon thin films were deposited with the second hollow electrode configuration, which presents the best performance as a high density source, according to the electrical parameters above. The deposition rate was examined as a function of different parameters and at diverse conditions. Initially, the deposition rate was measured for different total gas pressures while the other plasma parameters were kept constant. Figure 7 shows the variation of the film growth rate as a function of the total gas pressure and there is an optimum at 200 Pa, where DR has a value of 0.76 nm/s.

The deposition rate is affected by almost all experimental conditions and its maximum can be shifted for other conditions. Nevertheless, the hollow electrode configuration, except for the relatively high electron densities that derived from the electrical measurements, presents also higher deposition rate at relatively low pressure. Moreover, the effect of the discharge power and the way of SiH_4 introduction to the discharge (through hollow cathode or not) on the deposition rate were examined under conditions of constant total gas pressure for the HC2 configuration. The pressure was set at 133 Pa, the electrode gap at 1.5 cm and the mixture of SiH_4 in H_2 at 10 %. Figure 8 shows that the deposition rate increases with the discharge power for all configurations. The HC2 mode allows higher energy transfer from the generator to the plasma for the same applied voltage, leading thus to much higher film growth rates compared to the CC configuration, especially in the case that SiH_4 is introduced outside the hollow. The silane in this case reaches the plasma region outside the cavity, where the electrons and H atoms density is relatively high, due to the enhanced ionization and dissociation in this area. The rate increases with the discharge power and presents a maximum of 0.81 nm/s at 44 W power dissipation in the discharge. This enhancement can be attributed to the increase of electron density and the number of electrons with enough energy to induce SiH_4 dissociation towards radicals, which are responsible for the deposition rate. On the other hand, the deposition rate increase is not so pronounced when the silane is fed through the hollow cathode, despite the high discharge power. In this case, the silane dissociation inside the hollow cathode or the preferential ionization of SiH_4 inside the cavities that consume SiH_4 without contributing to the film growth may limit the deposition rate.

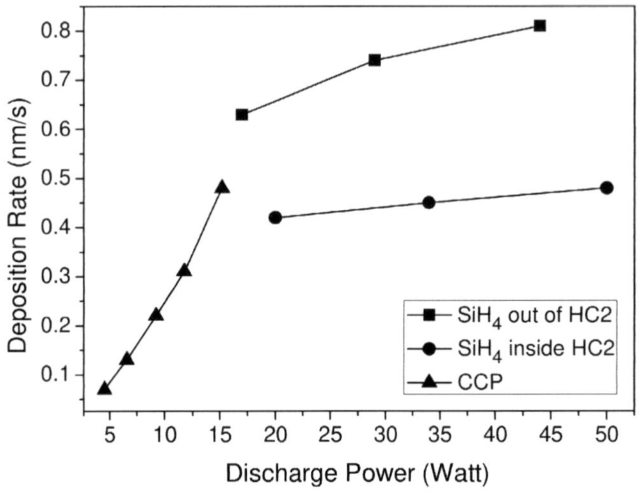

Figure 8. The deposition rate as a function of the discharge power at 1.5 cm distance and for two different sources and two different ways of feeding the silane in the reactor.

4. Conclusions

Two different hollow cathode configurations have been constructed and their operation has been tested and compared to a typical CCP source by using electrical measurements. The hollow configuration of wider cavities had the best performance as far as the electron density is concerned. Electron densities of above 10^9 cm^{-3} were achieved with the hollow cathode, much higher than the typical Capacitively Coupled discharges.

Microcrystalline silicon thin films depositions were also performed and the effect of the different geometries and sources on the film growth rate was investigated. The hollow cathode configuration was found that allows higher energy transfer from the generator to the plasma as compared to the CC source and this leads to a significant enhancement of the film growth. The introduction of SiH_4 directly into the plasma zone instead of through the hollow cathode had also beneficial effect on the deposition rate. Further work is required for optimizing the geometry of the HC source and the gas mass transfer in order to obtain the optimum results on both the film growth rate and the film quality.

References

[1] Meier J, Flückiger R, Keppner H and Shah A 1994 *Appl. Phys. Lett.* **65**

[2] Meier J, Dubail S, Cuperus J, Kroll U, Platz R, Torres P, Anna Selvan JA, Pernet P, Beck N, Pellaton Vaucher N, Hof Ch, Fischer D, Keppner H and Shah A 1998 *J. Non-Cryst. Solids* **227-230** 1250

[3] Feitknecht L, Kluth O, Ziegler Y, Niquille X, Torres P, Meier J, Wyrsh N and Shah A 2001 *Sol. Energy Mater. Sol. Cells* **66** 397

[4] Vetterl O, Lambertz A, Dasgupta A, Finger F, Rech B, Kluth O and Wagner H 2001 *Sol. Energy Mater. Sol. Cells* **66** 345

[5] Guo L, Kondo M, Fukawa M, Saitoh K and Matsuda A 1998 *Jpn. J. Appl. Phys.* **37** L1116 Part2

[6] Fukawa M, Suzuki S, Guo L, Kondo M and Matsuda A 2001 *Sol. Energy Mater. Sol. Cells* **66** 217

[7] Roschek T, Repmann T, Müller J, Rech B and Wagner H 2002 *J. Vac. Sci. Technol.* A **20** 492

[8] Jia H, Saha JK, Ohse N and Shirai H 2007 *Thin Solid Films* **515** 6713
[9] Jia H, Kuraseko H and Kondo M 2008 *J. Non-Cryst. Solids* **354** 2083
[10] Wu JH, Shieh JM, Dai BT and Wu WS 2004 *Electrochem. Solid-State Lett.* **7** G128
[11] Kosku N and Miyazaki S 2006 *Thin Solid Films* **511** 265
[12] Thang DH, Muta H and Kawai Y 2008 *Thin Solid Films* **516** 4452
[13] Roca i Cabarrocas P, Bulkin P, Daineka D, Dao TH, Leempoel P, Descamps P, Kervyn de Meerendre T and Charliac J 2008 *Thin Solid Films* **516** 6834
[14] Ohtsu Y and Fujita H 2008 *Appl. Phys. Lett.* **92** 171501
[15] Bardos L 1996 *Surf. Coat. Technol.* **86-87** 648
[16] Bardos L, Barankova H and Berg S 1997 *Surf. Coat. Technol.* **97** 723
[17] Barankova H and Bardos L 1999 *Surf. Coat. Technol.* **120-121** 704
[18] Tabuchi T, Takashiri M and Mizukami H 2003 *Surf. Coat. Technol.* **173** 243
[19] Niikura C, Kondo M and Matsuda A 2004 *J. Non-Cryst. Solids* **338-340** 42
[20] Lee YS, Lee HS and Chang HY 2010 *Thin Solid Films* in press
[21] Niikura C, Itagaki N, Kondo M, Kawai Y and Matsuda A 2004 *Thin Solid Films* **457** 84
[22] Kocka J, Mates T, Ledinsky M, Stuchlikova H, Stuchlik J and Fejfar A 2008 *Thin Solid Films* **516** 4966
[23] Spiliopoulos N, Mataras D and Rapakoulias DE 1996 *J. Vac. Sci. Technol.* A **14**
[24] Amanatides E and Mataras D 2001 *J. Appl. Phys.* **89** 1556
[25] Amanatides E, Hammad A, Katsia E and Mataras D 2005 *J. Appl. Phys.* **97** 073303
[26] Godyak VA, Piejak RB and Alexandrovich BM 1991 *J. Appl. Phys.* **69** 3455

Journal of Physics: Conference Series **275** (2011) 012015 doi:10.1088/1742-6596/275/1/012015

Plasma gasification of waste as a method of energy saving

V E Popov, A N Bratsev, V A Kuznetsov, S V Shtengel and A A Ufimtsev
Institute for Electrophysics and Electric Power RAS
18 Dvortsovaya nab., St.-Petersburg, 191186, Russia

E-mail: lab10@iperas.nw.ru

Abstract. Several versions of the organizations of the process of plasma-chemical gasification with the use of air, carbon dioxide, steam and their mixtures as the plasma-forming gas are considered in the presentation. The results of the calculation-theoretical evaluations of the quality of synthesis gas and efficiency of gasification, and also the results of experiments on plasma gasification of wood waste carried out on the experimental IEE RAS test-bench are given. The results of calculations are compared with experimental data.

1. Introduction

In the second half of the 20-th centuries the rate of waste formation and energy consumption per capita steadily grew together with increase of a life standard. In 2008 capacity of global consumption of primary energy made approximately $4.73{\cdot}10^{11}$ GJ [1] at total population size of ~$6.68{\cdot}10^{9}$ person [2]. That is, on average one person consumes ~2.24 kW of primary energy. In the developed countries 400-760 kg of municipal waste are formed annually per capita [3] and it is consumed 60-390 GJ per capita [1]. If to assume, that combustion heat of municipal solid waste (MSW) makes 15-20 MJ/kg, then the use of waste will allow saving about 3-15 % of primary energy. These wastes on 15-68 % consist of paper and cardboard which composition is similar to wood. Besides, wood materials in their self are the important source of renewed energy [4].

The process of direct burning of biomass and waste for energy generation is ineffective [5] and ecologically hazardous. Now gaseous fuel combustion has higher efficiency. Therefore, it is expedient to subject solid fuel to the gasification process as a result of which primary energy of solid fuel will be transformed to chemical energy (energy of combustion) of synthesis gas (mixture of hydrogen and carbon monoxide). Besides, the produced synthesis gas can be a raw material for production of liquid fuel [6] and hydrogen.

Due to plasma application it is possible to increase power efficiency of gasification and to decrease the content of incombustible components of synthesis gas (nitrogen, carbon dioxide). Thus energy content of the plasma gasification products can exceed the combustion heat of initial raw material. Besides, it allows decreasing of the specific emission of greenhouse gases both on power stations, and at incineration plants.

Synthesis gas can be used as a raw material for production of hydrogen, liquid fuel and other substances of chemical-technological purpose. Synthesis gas can be produced from hydrocarbonic waste of various morphological compositions [7].

2. Ways of organization of plasma gasification and their features

There are several ways of solid fuel gasification, namely: updraft, downdraft, twin-fire, crossdraft, entrained flow, fluidized bed. These schemes can be transferred without change on plasma gasification methods. Figure 1 shows inlet and outlet points of mass flows for various schemes, and also characteristic zones of the gasification processes.

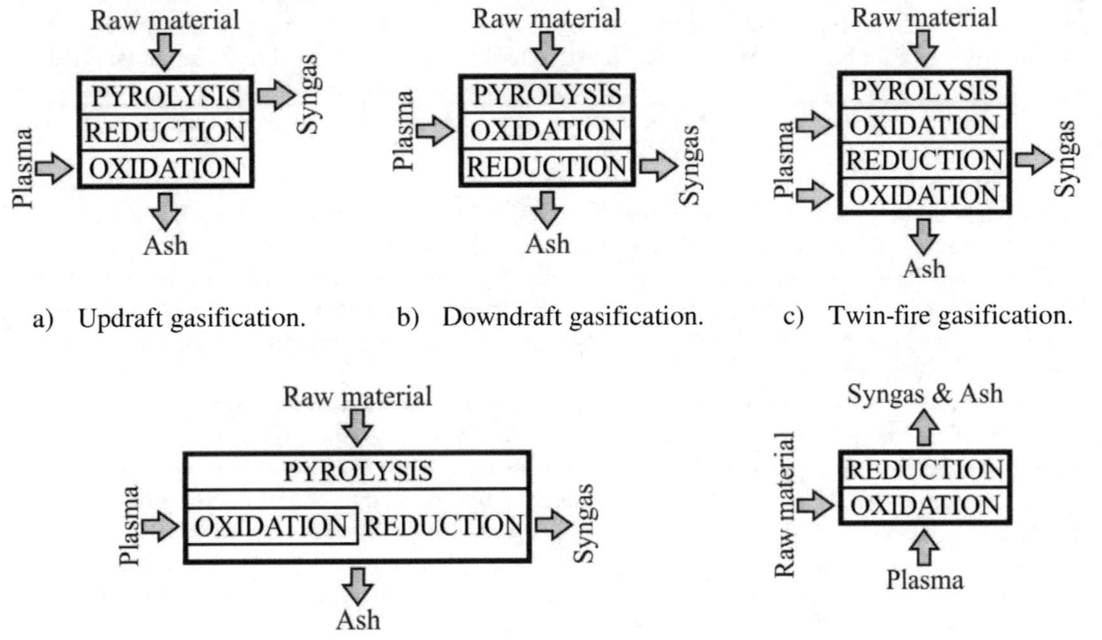

a) Updraft gasification. b) Downdraft gasification. c) Twin-fire gasification.

d) Crossdraft gasification. e) Entrained flow gasification.

Figure 1. Location of the characteristic zones and inlet and outlet points of mass flows for various ways of organization of plasma gasification.

Two basic zones: oxidizing and reduction can be marked out in all schemes. The oxidizing zone is located at a point of plasma inlet, therefore the highest temperatures are observed in it and the most part of chemical reactions proceeds. The reduction zone is next to the oxidizing zone (in the direction of gas movement). There the oxidation of semicoking products by carbon dioxide and steam takes place. If fuel is supplied into the low temperature zone, so it is possible to separate a zone of pyrolysis in the reactor volume. Temperatures in this zone are high enough for evaporation of moisture and primary decomposition of fuel with release of simple gases and tars, but insufficient for reduction reactions. There is no oxygen in this zone; therefore the composition of products of pyrolysis of raw material is determined only by the rate of heating and pressure.

Updraft plasma gasification is well adapted for high humidity biomass gasification due to availability of thermal energy of syngas for raw material drying. The hot gases, ascending from the zone of oxidizer inlet, heating up cold biomass, intensifying pyrolysis and evaporation of an excess moisture. The evaporated moisture does not get into zones of oxidation and reduction, therefore less energy is required for its heating. However, products of low-temperature pyrolysis do not pass through high-temperature zones and as a result the synthesis gas is polluted by tars. There are solid products of pyrolysis having high combustion heat in a zone of plasma inlet. Therefore, high temperatures are achieved in this zone even at use of air blasting without energy input. For example, at gasification of graphite by air in the adiabatic mode the temperature of products (in which CO_2 practically is absent) is ~1200 °C. Use of air plasma in this case will lead to increase in temperature in the oxidizing zone, to increase the reduction zone and insignificant reduction in the amount of incombustible admixtures in synthesis gas. Use of CO_2 or H_2O as a plasma forming gas will lead to increase in a specific energy output from a mass unit of raw material and to increase in power inputs. The important advantage of the method is the opportunity of liquid slag discharge.

In the downdraft gasification process all products of pyrolysis pass through the high temperature zone in which the tars are subjected to cracking. That is why the syngas at the gasifier outlet contains insignificant amount of tar (comparing with the previous method). Then, steam and carbon dioxide take part in the oxidation process of solid products of the pyrolysis in the reduction zone. Reactions of CO_2 and H_2O with carbon are exothermal that is why high temperature should be maintained in the reduction zone. Thus, input of energy with plasma will lead to intensification of reduction reactions and to deeper tar cracking. It will result in increase in power efficiency of gasification and decrease in the content of incombustible admixtures in syngas.

Twin-fire gasification is an improved version of the previous one. At the expense of the additional point of plasma inlet the temperature in the reduction zone increases and its length can be increased. Solid products of pyrolysis are in the lower oxidizing zone with the increased ash content, as a part of their carbon has been gasified in the reduction zone by the products of plasma conversion of volatile substances. That is, the basic purpose of this zone is increase of a temperature level in the reduction zone due to burning out of residual carbon. By analogy with the updraft process, it is possible to organize liquid slag discharge.

In crossdraft gasification the point of plasma inlet is at the same level with the point of gas outlet. The oxidizing and reduction zones divide the reactor shaft in horizontal plane. That is why part of the oxidizing products gets into the oxidizing zone and is subjected to deep cracking, while other part gets in the reduction zone. As a rule, these zones are high-temperature therefore the content of tar in synthesis gas will be less than in the updraft process. However, because of small length of the reduction zone the content of H_2O and CO_2 in raw synthesis gas will be higher than in the downdraft process. This method differs in the increased temperatures of synthesis gas at the reactor outlet.

In entrained flow gasification the speeds and trajectories of movement of solid and gas phases coincide. The release of volatile agents and water evaporation happens directly in the oxidation zone. Residence time of the solid phase in the reduction zone is short in comparison with other ways. Shredding of raw material provides the increase in level conversion of raw material, and supply of the excess oxidizer amount provides high temperatures. The energy efficiency of this method in comparison with the downdraft process will be lower due to higher energy consumption for achieving of higher temperatures. Besides, synthesis gas will be highly polluted with solid particles. Application of plasma allows achievement of the same temperatures at the decreased oxidizer consumption. The mixing (initial) temperature will be considerably higher than the average temperature in the reactor. Recirculation of solid particles is one more way of increase in conversion level of raw material. In the case of low reactivity of fuel the recirculation use results in increase of power inputs.

The general flow diagram of the process of fluidized bed gasification, except for position of a point of raw material inlet, corresponds to gasification in the entrained flow (figure 1 e). The reaction volume of gasifiers working under this scheme as a rule has a shape expanding to the top. In the top layers of fluidized bed there are the smallest pieces of raw material, and in the bottom layers are the biggest ones. Constant mixing of pieces of fuel promotes more uniform distribution of temperatures in cross-section section of the reactor. Depending on the location of the point of raw material inlet this method will have similar characteristics either with the updraft or with the downdraft process. If the raw material is supplied from the top, so during falling and migrations of a piece in a high-temperature bottom layer it will have time to lose a part of the moisture and volatile agents that results in synthesis gas pollution with tar. If the raw material is entered from the bottom, an output of volatile agents and evaporation of water is carried out in the oxidizing zone where tars are subjected to deep cracking. As well as at gasification in the entrained flow the synthesis gas will be highly polluted by solid particles. Their size will be determined by the least speed of gas flow at the top part of the gasifier.

Thus, energy of plasma will be most effectively used in the downdraft and twin-fire processes (under condition of supply of the basic plasma stream in the top inlet point). It is caused by mixture of plasma with pyrolysis products directly in the inlet plasma point and long residence time of solid

Journal of Physics: Conference Series **275** (2011) 012015 doi:10.1088/1742-6596/275/1/012015

components in a high-temperature zone. Use of steam or CO_2 plasma can also be justified in the updraft process. For example, if it is necessary to produce high-calorific synthesis gas from fuels with high humidity.

3. The description of a design procedure

The developers of technologies of synthesis gas use are greatly interested in its composition, power inputs and specific yield of chemical energy of gas.

The following assumptions have been made for modeling of the plasma gasification process and calculation of potential parameters of processes of electric power generation and synthesis of liquid fuels:

- the mixture of plasma and ash-free mass of wood waste reaches the thermodynamic equilibrium state at temperature of 1500 K and pressure of 101.3 kPa;
- the inorganic component of wood waste does not participate in chemical reactions, and power inputs on its heating is from 298 K up to 1500 K that is ~1.367 MJ/kg;
- there are no changes in composition of gasification products at their cooling to 298 K;
- mass composition of wood waste: C – 50.25, H – 6.09, O – 43.35, N – 0.2, S – 0.1 % (dry, ash-free mass), ash content– 1 % (dry mass), humidity – 20 % (as received); LHV – 13.90 MJ/kg (as received) [8];
- mass composition of air: N_2 – 74.43, O_2 – 22.81, Ar – 1.27, CO_2 – 0.04, H_2O – 1.45 %;
- dependence between efficiency of combined cycle and temperature in range of 1400-2000 K, as it is described in [9], was approximated by polynomial to estimate specific yield of electric energy:

$$\eta = 0.133 + 4.3\cdot10^{-5}\cdot T + 2.6\cdot10^{-7}\cdot T^2 - 7.9\cdot10^{-11}\cdot T^3 \quad (1)$$

  where η – efficiency of combined cycle (part of an unit), T – temperature (K);
- efficiency of combined cycle at temperatures higher then 2000K was considered as constant and equal to efficiency at 2000K;
- if volumetric ratio $H_2/CO \geq 2$ then it is potentially possible (in conformity with stoichiometry of Fischer-Tropsch process), that the specific yield of liquid fuel is defined by the formula:

$$G_{SF} = G_{SG,d}\cdot\chi_{CO}\cdot\rho_{CH2} \quad (2)$$

  where G_{SF} – liquid fuel yield (kg/kg), $G_{SG,d}$ – dry syngas yield (Nm^3/kg), χ_{CO} – volume fraction of CO in syngas (part of an unit), ρ_{CH2} – density (at 25 °C, 101.3 kPa) of the conceptual gas with chemical formula CH_2 (kg/m^3);
- if volumetric ratio $H_2/CO < 2$ then it is potentially possible (in conformity with stoichiometry of Fisher-Tropsh process), that the specific yield of liquid fuel is defined by the formula:

$$G_{SF} = G_{SG,d}\cdot(\chi_{H2}+\chi_{CO})\cdot\rho_{CH2}/3 \quad (3)$$

  where χ_{H2} – volume fraction of H_2 in syngas (part of an unit).

4. Results of calculations of synthesis gas composition and efficiency of plasma gasification of wood waste

Figure 2 represents the dependence of composition of dry syngas versus specific power inputs (per mass unit of wood waste) at use of air, carbon dioxide and steam as a plasma forming environments.

It is obvious from (figure 2) that at wood waste gasification by air plasma the increase in the specific power inputs (on 1 kg of raw material) leads to the increase in concentration of combustible components (H_2 and CO) and decrease in concentration of incombustible gases (CO_2 and N_2). The reverse tendency is observed at use of carbon dioxide or steam. It is caused by the energy of plasma of "passive" oxidizers (H_2O or CO_2) goes not only on a covering of endothermic effect of gasification, but also for decomposition of molecules of an oxidizer, and its flowrate increases with growth of power inputs for passive oxidizers (see figure 3).

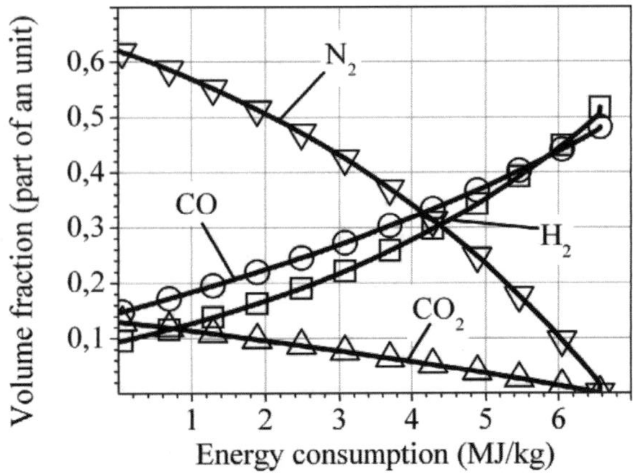

a) Air-plasma gasification.

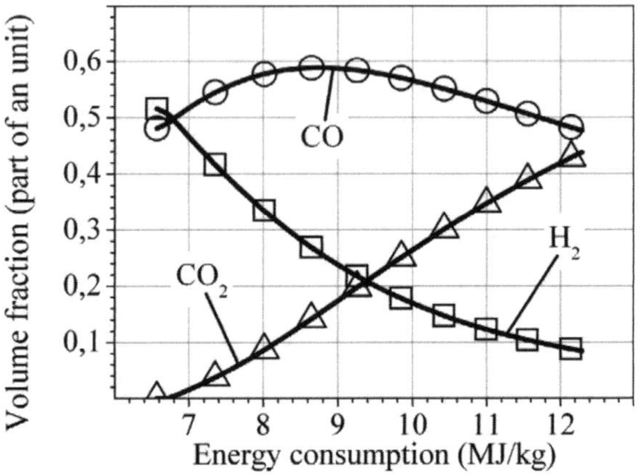

b) CO_2-plasma gasification.

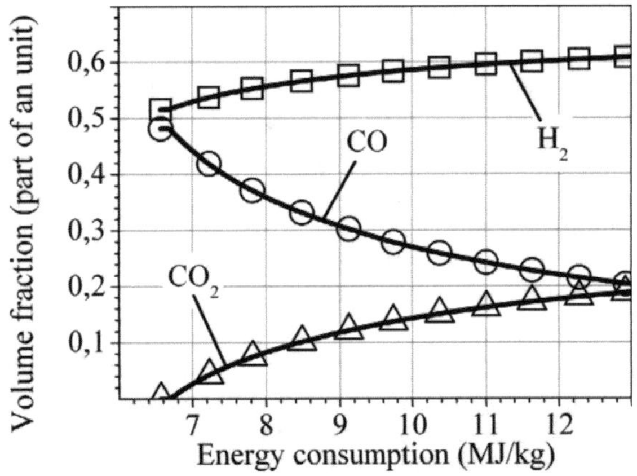

c) Steam-plasma gasification.

Figure 2. Influence of energy consumption on volumetric composition of dry gas at plasma gasification of wood waste.

Relation H_2/CO over the calculation range changes from 0.64 to1.07 at use of air plasma, 0.18-1.07 for CO_2 plasma, 1.07-3.65 for steam plasma. Presence of the relation 1.07 common for all is explained by existence of the common mode of plasma gasification for all oxidizers – a mode of pyrolysis (if the oxidizer flowrate vanish, parameters of the process aspire to the pyrolysis parameters).

Differences in tendencies of component concentration variations are caused by the character of changes introduced by the oxidizer in the element balance of a mixture of fuel with an oxidizer. At increase in the air flowrate the percentage of nitrogen and oxygen in the mixture increase, and relation of carbon and hydrogen weakly varies. At increase in CO_2 flowrate the percentage of oxygen and carbon increase. At increase in the steam flowrate the percentage of oxygen and hydrogen increase. At plasma gasification it is possible to vary the relation H_2/CO over a wide range, changing the plasma forming gas composition and compensating the energy consumption changes for the process due to energy input from the outside.

At air-plasma gasification the yield of chemical energy linearly increases with increase in energy consumption (figure 4): on 1 MJ of the input energy the chemical energy yield of synthesis gas increases on ~1.55 MJ. Energy consumption for pyrolysis of wood make ~6.57 MJ/kg. The maximum of specific chemical energy yield is reached near to the point of pyrolysis.

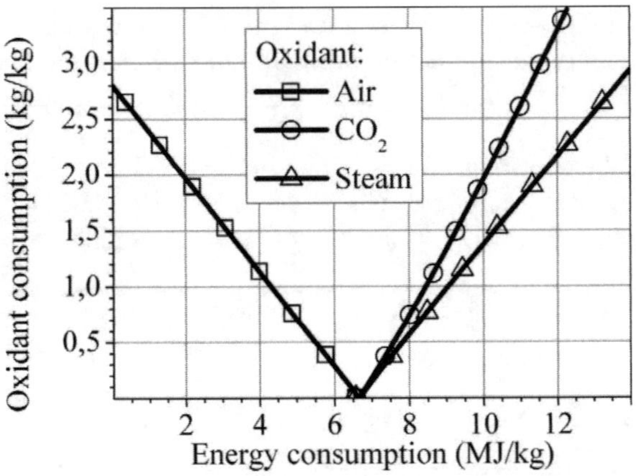

Figure 3. Dependence of the specific oxidizer flowrate versus energy consumption.

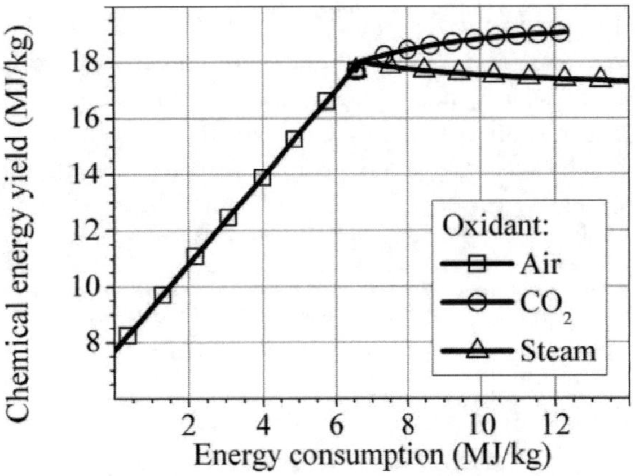

Figure 4. Dependence of the specific chemical energy yield versus energy consumption.

At use of CO_2 plasma the specific chemical energy yield grows with increase in power inputs, and for steam plasma – it decreases. It is caused by displacement of equilibrium state (at 1500 K) aside to hydrogen oxidations and carbon dioxide reduction.

Figure 5 shows the estimation of influence of the specific energy consumption on the specific yield of electric power from the system using plasma gasification and the combined cycle for the electric power generation from wood waste. It is visible that use of air plasma allows increasing on ~20 % of the electric power yield from a mass unit of raw material, and use of steam and carbon dioxide plasma is not expedient.

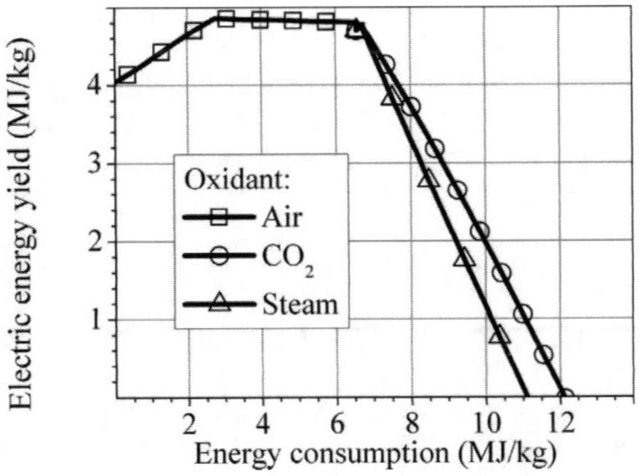

Figure 5. Dependence of the specific yield of the net electric energy versus energy consumption.

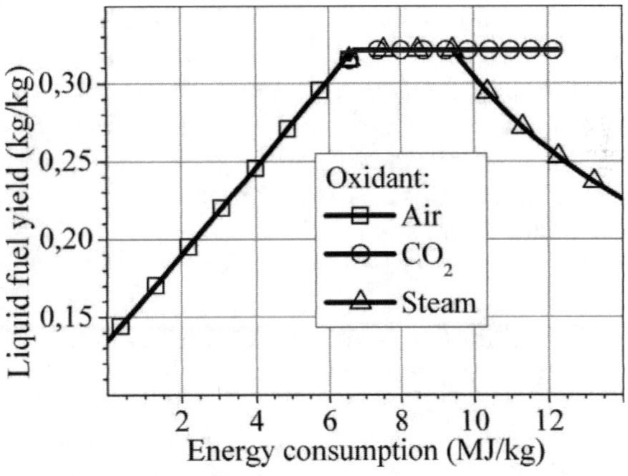

Figure 6. Dependence of the specific yield of the liquid fuel versus energy consumption.

Figure 6 represents the estimation of influence of specific energy consumption on the specific yield of synthetic liquid fuel from the system using plasma gasification and Fischer-Tropsch process for liquid fuel production from wood waste. It is visible that use of air plasma allows increasing the yield of liquid fuel from a mass unit of raw material more than in 2 times, and use of steam and carbon dioxide plasma leads to rise in price of the end-product.

5. Experiment results on air-plasma gasification of wood

The pilot installation for plasma gasification of carbonaceous materials was created in IEE RAS [10]. On this installation a series of experiments on plasma gasification of wood by air has been performed

[11,12]. In the beginning of each experiment the installation was warmed up within 6-8 hours due to gasification of charcoal. Charcoal was loaded periodically during the warming up phase. After that the wood was loaded periodically instead of charcoal and the reactor switched over its mode to plasma gasification of wood process within 1-2 hours.

The wood with moisture of ~20 % (measured by the continuous drying method) and LHV of ~13.9 MJ/kg was used in the tests. It was assumed that dry wood mass consists of C - 49.75, H - 6.03, O - 42.92, N - 0.20, S - 0.10, ash – 1 %, mass composition of blowing in air is N_2 – 74.4, O_2 – 22.8, Ar – 1.27, CO_2 – 0.04, H_2O – 1.45 %.

In table 1 the parameters of two characteristic experimental modes of air-plasma gasification of wood are compared with calculated values.

Table 1. Comparison of experimental and calculated data on air-plasma gasification of wood.

Source	E_{in} [a] (MJ/kg)	G_{in} [b] (kg/kg)	$G_{SG,d}$ (Nm^3/kg)	Composition of dry syngas (%vol.) H_2	CO	N_2	O_2	Ar	CO_2	Q_{out} [c] (MJ/kg)
exp.	3.10	1.22	2.38	32.2	26.1	33.4	0.27	0.40	7.64	14.7
calc.	3.10	1.22	2.22	27.2	29.3	35.8		0.43	7.22	13.5
exp.	2.16	1.44	2.45	28.0	23.6	38.2	0.28	0.46	9.30	13.5
calc.	2.16	1.44	2.34	24.7	25.6	40.1		0.48	9.01	12.7

[a] Specific energy consumption.
[b] Specific air consumption.
[c] Specific yield of the syngas chemical energy.

The same deviations of concentration of hydrogen and carbon monoxide in experimental and calculated data are explained by effect of water-gas shift reaction at cooling of synthesis gas. It leads to reduction in CO concentration and increase in H_2 and CO_2 concentration in dry synthesis gas. This effect is not considered in calculation. The increase in experimental (in comparison with calculated) values of a specific yield of chemical energy of synthesis gas happens due to the temperature field influence. Stationary temperature of products of plasma gasification of charcoal is higher than the similar parameter for wood. Therefore reactor lining is warmed up to higher temperatures, than at plasma gasification of wood. At transition from charcoal to wood the lining starts to give a part of the saved energy to the gasification products. As a whole the results of experiments are well agreed with calculated data.

6. Conclusion

The downdraft and twin-fire schemes of gasification allow most effectively use of plasma energy due to mixture of pyrolysis products with plasma in a point of its inlet and long residence time of solid pyrolysis products in a high-temperature zone. The most suitable oxidizer for plasma gasification of wood waste is air as on 1 MJ of the input energy the chemical energy yield of synthesis gas increases on ~1.55 MJ. Application of plasma during gasification allows increasing the efficiency of the electric power generation and liquid fuel production from wood waste.

References

[1] 2009 *BP Statistical Review of World Energy* BP plc. 57th edition. 45 p.
[2] "Total Midyear Population for the World" - http://www.census.gov/ipc/www/idb/worldpop.php (accessed: 1 June 2010)
[3] "OECD Environmental Data Compendium" – http://www.oecd.org/dataoecd/22/59/41878217.xls (accessed: 1 June 2010)
[4] Huber G W and Dale B E 2009 Grassoline at the Pump. *Sci. Am.* **301** 52–9
[5] Patumsawad S 2002 Supplementing coal with MSW in a fluidised bed combustor. *The journal of KMITNB* **12** 19–25
[6] Bratsev A N, Kuznetsov V A, Popov V E, Rutberg A Ph, Ufimtsev A A and Shtengel S V 2009

Journal of Physics: Conference Series **275** (2011) 012015 doi:10.1088/1742-6596/275/1/012015

Experimental development of methods on plasma gasification of coal as the basis for creation of liquid fuel technology. *High Temp. Mater. Processes* **13** 147–54

[7] Yun Y, Chung S W and Yoo Y D 2003 Syngas quality in gasification of high moisture municipal solid wastes. *Prepr. Pap.-Am. Chem. Soc., Div. Fuel Chem.*, 48(2), 823–4

[8] 2007 *Perry's Chemical Engineers' Handbook* Ed D W Green and R H Perry. United States. McGraw-Hill. Eighth Edition. 2400 p.

[9] Ishikawa M, Terauchi M, Komori T and Yasuraoka J 2008 Development of high efficiency gas turbine combined cycle power plant. *Mitsubishi heavy industries, Ltd. Technical Review* **45** 15–7

[10] Bratsev A N, Popov V E, Shtengel S V and Rutberg A Ph. 2006 Some aspects of development and creation of plasma technology for solid waste gasification. *High Temp. Mater. Processes* **10** 549–56

[11] Bratsev A N, Glezin I L, Kovshechnikov V B, Kumkova I I, Kuznetsov V A, Popov V E, Shtengel S V and Ufimtsev A A 2007 Experimental research of air gasification of waste. The first results. *XXVIII International Conference on Phenomena in Ionized Gases (ICPIG 2007)* Czech Republic, Prague. Institute of Plasma Physics. p. 1848–51

[12] Kuznetsov V A, Bratsev A N, Kovshechnikov V B, Kumkova I I, Popov V E, Shtengel S V and Ufimtsev A A 2007 Distinctive features of biomass gasification using ac plasma generators working on air. *IEEE Pulsed Power Conference (PPPS-2007).* USA, New Mexico, Albuquerque. Omnipress, Madison USA. p. 1223–6

Journal of Physics: Conference Series **275** (2011) 012016 doi:10.1088/1742-6596/275/1/012016

CO_2 valorization by means of Dielectric Barrier Discharge

H Machrafi[1,*], S Cavadias[2], J Amouroux[2]
[1]Thermodynamics of Irreversible Phenomena, Allée du Six-Août, 17, B-4000 Liège, Belgium
[2]Laboratoire de Génie des Procédés Plasmas et Traitements de Surfaces, 11, rue Pierre et Marie Curie, F-75005 Paris, France

E-mail: H.Machrafi@ulg.ac.be

Abstract. As atmospheric pollution is causing several environmental problems it is incumbent to reduce the impact of pollution on the environment. One particular problem is the production of CO_2 by many transport and industrial applications. Instead of stocking CO_2 and instead of being a product, it can be used as a source. The case considered is the CO_2 reformation of methane producing hydrogen and CO. It is an endothermic reaction, for which the activation barrier needs to be surpassed. This can be done efficiently by the method of Dielectric Barrier Discharge. The process relies on the collision of electrons, which are accelerated under an electrical field that is created in the discharge area. This leads to the formation of reactive species, which facilitate the abovementioned reaction. This study is performed using a Matlab program with the Reaction Engineering module in COMSOL (with an incorporated kinetic mechanism) in order to model the discharge phase. Then COMSOL (continuity and Navier-Stokes equations) is used to model the flow in the post-discharge phase. The results showed that both a 2D and 3D model can be used to model the chemical-plasma process. These methods need strongly reduced kinetic mechanism, which in some cases can cause loss of precision.

1. Introduction

Reducing the pollution that is caused by industry and transport has been investigated widely and a part of many efforts that have been made is spent for reducing the emission of CO_2. Of many ways, depending on the applications, this paper considers CO_2 as a main source in order to reform methane by means of Dielectric Barrier Discharge (DBD). This implies that CO_2 is thus used as a source product instead of being rejected into the atmosphere. The study concerns modeling work that is inspired by an experimental set-up (prepared in the laboratory LGPPTS), which is presented in figure 1. The length of the reactor is 120 mm with a diameter of 30 mm, having a 2 mm thick glass wall. The high voltage electrode is 2 mm thick and has a diameter of 26 mm, being consisted out of stainless steel. A gap of 2 mm is allowed between the glass wall and the electrode. The electrode that stands for the ground electrode is a wire sheet of 20 mm width and 2 mm thick, wrapped around the glass tube at the height level of the electrode as is depicted in figure 1. A peak to peak alternative voltage from 5 to 30 kV is supplied by a high voltage generator. The discharge is filamentary. This experimental set-up is taken as the basis for the modeling.

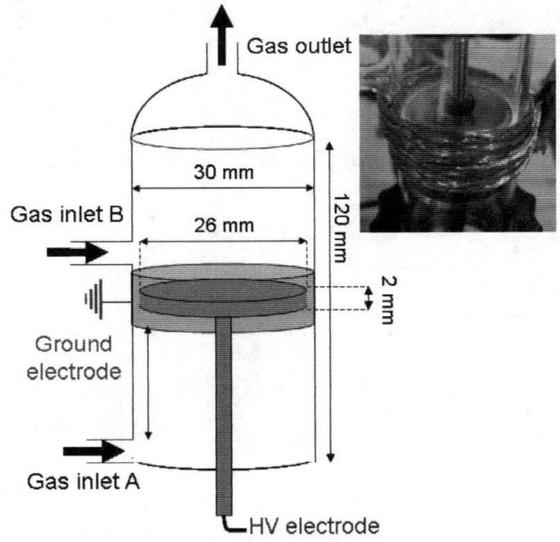

Figure 1: The DBD reactor

The overall reaction that is of interest is given by
$CO_2 + CH_4 \rightarrow 2\ CO + 2\ H_2 \quad \Delta H^{\circ} = 258.9$ kJ/mole.
This, also called the CO_2 reforming of methane, is an endothermic reaction, needing energy to surpass its activation barrier. This energy is provided by electrons that are energetic enough to initiate this process by decomposing CO_2 into CO and O radicals, as is also suggested by [1] and used in [2]:
$CO_2 + e^- \rightarrow CO + O\bullet + e^-$
This initiation reaction is also considered by [3]. The following reaction is also proposed by [1]:
$CO_2 + e^- \rightarrow C + 2O\bullet + e^-$
The carbon atom is observed to deposit on to the electrode surfaces, therefore leaving only electrons and oxygen radicals for the reformation of methane into H_2. This reaction sequence is proposed by [1], supported by experiments in a plasma environment, to proceed as follows:
$CH_4 + O\bullet \rightarrow \bullet CH_3 + \bullet OH$
$\bullet CH_3 + O\bullet \rightarrow H_2CO + H\bullet$
The H_2CO is unstable in cold plasma and gives finally CO and H_2:
$H_2CO + (e^-/O\bullet/H\bullet/\bullet OH) \rightarrow HCO\bullet + (H\bullet/\bullet OH/H_2/H_2O)$
$HCO\bullet + (H\bullet/\bullet O/\bullet OH\ /\bullet CH_3) \rightarrow CO + (H_2/\bullet OH/H_2O/CH_4)$

Having set out the main kinetic pathways, this work can be divided in four steps. Before, introducing these steps, the idea is to introduce CO_2 at the bottom of the reactor through Gas Inlet A (figure 1) and CH_4 through Gas Inlet B. The first step is a 3D COMSOL model, with only the Navier-Stokes equations, in order to determine the velocity fields. A convection-diffusion model is also added in order to look to the diffusion characteristics and to see whether the carbon could have the possibility to deposit on the electrodes. The second step is to use a 2D model with the complete kinetic model (based on the kinetic pathways, discussed earlier and shown later on), so that the concentrations before entering the post-discharge region can be estimated. Since only few reactions can be considered for the 3D model, for computing memory limits, the third step is to reduce the kinetic model having a reasonable agreement with the initial one, using Reaction Engineering Lab in COMSOL. The last step is to use the 3D model, with only the post-discharge region, having the concentrations of the essential species that are calculated by the 2D model.

Journal of Physics: Conference Series **275** (2011) 012016 doi:10.1088/1742-6596/275/1/012016

2. Hydrodynamics, convection and diffusion in the DBD reactor

2.1 Modeling flow in the 3D reactor

The 3D model as is used in the modeling work is presented in figure 2 with the subdomains, boundaries and mesh.

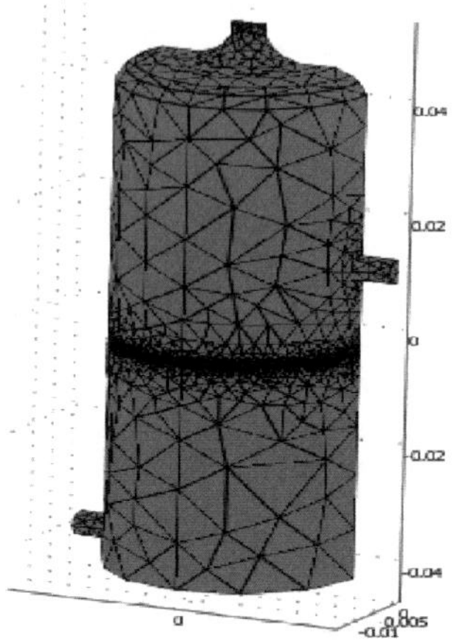

Figure 2: The mesh of the DBD reactor model

CO_2 is let trough at the bottom entrance of the reactor and goes through the discharge region between the reactor wall and the high voltage electrode. The CO_2 flow then encounters the CH_4 flow that is allowed to enter at the top entrance of the reactor (to the right), 10 mm above the electrode in the post-discharge area. The flow rates for CO_2 and CH_4 are 1 *l*/min. This corresponds to respectively 1.3 m/s and 1 m/s (normal to the boundary). The velocity and pressure fields are then calculated by the "Incompressible Navier Stokes" model (ns). Figure 3 shows a "slice" of the velocity field and streamlines at the level of the entrances. The value that is corresponding to the discharge region (around the inlet stream) is not zero but equals around 0.021 m/s. The streamlines show in the discharge region a stable velocity value in contrary to the post-discharge region. This allows using the velocity 0.021 m/s as an inlet velocity for the 2D and 3D model reactors.

2.2 The convection and diffusion in the 3D reactor

The purpose is to find out the convection and diffusion behavior of the two basic constituents CO_2 and CH_4 in the geometry of the present reactor. A convection and diffusion model (chcd) is added to the "ns" model. A 1:1 ratio (CO_2:CH_4) between the concentrations of these components is chosen of $1 mol/m^3$. Figure 4 shows the concentrations of CO_2 and CH_4 obtained from solving for the convection-diffusion equations for these components. Figure 4 shows about the same diffusion characteristics for the two constituents. The deposit of carbon on the electrode would not be inevitable in these circumstances.

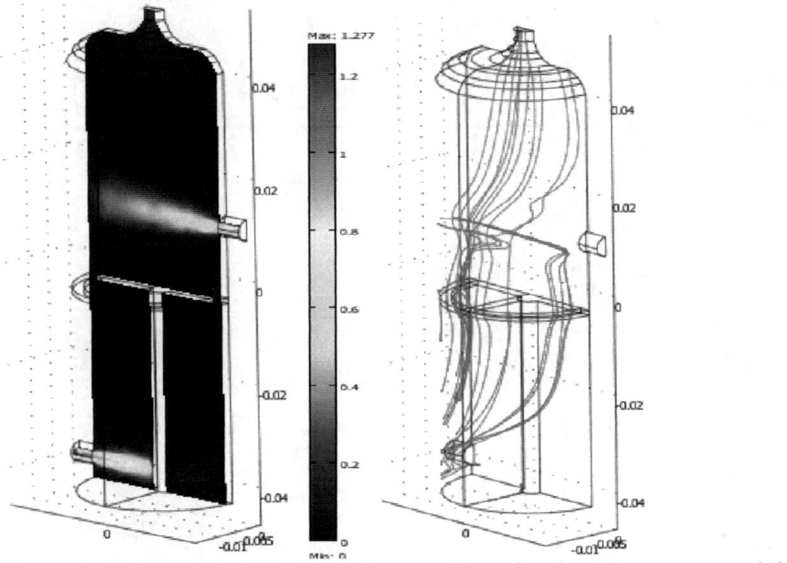

Figure 3: Velocity field (m/s) and streamlines for the 3D reactor model

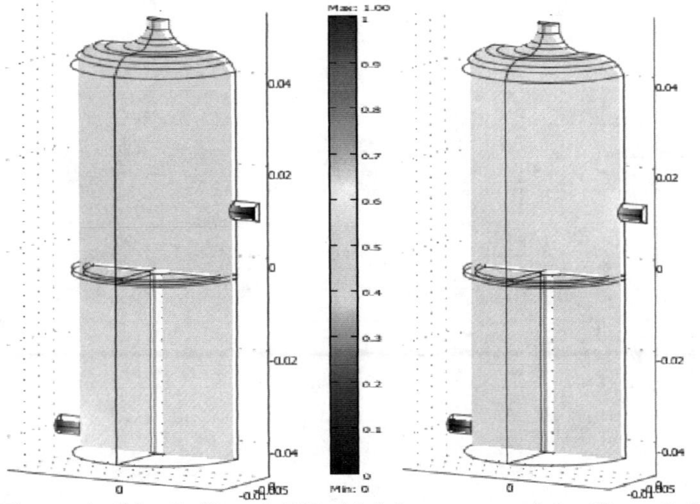

Figure 4: CO_2 (left) and CH_4 (right) concentrations (from 0 (blue) to 1 (red) mol/m^3) in the 3D reactor model

3. The 2D reactor model

3.1 The kinetic model

The kinetic model has been composed out of the reactions proposed in the introduction of which the reaction rates were found in the literature. Since the electrons could also decompose methane the following equations are also proposed [4,5]:

$CH_4 + e^- \rightarrow \bullet CH_3 + H\bullet + e^-$

$CH_4 + e^- \rightarrow \bullet CH_2 + H_2 + e^-$

$CH_4 + e^- \rightarrow \bullet CH_2 + 2H\bullet + e^-$

Journal of Physics: Conference Series **275** (2011) 012016 doi:10.1088/1742-6596/275/1/012016

$CH_4 + e^- \rightarrow \bullet CH + H\bullet + H_2 + e^-$
$CH_4 + e^- \rightarrow C + 2H_2 + e^-$
The deposition of carbon will be neglected and not considered in our case and it is assumed that the only electron interaction that is allowed concerns CO_2 and CH_4. Table 1 shows the kinetic model [5-7] in question. Further $CO_2/CO/O_2/O$ reactions [8] are included in table 1 describing the neutral interactions between these species. The kinetic mechanism contains 29 reactions and 15 species (T in [K], R in [kcal/(mol*K)]).

Table 1: Kinetic model of CO_2 reforming of methane (*Reaction rate in [cm^6 $molecule^{-1}$ s^{-1}]).

Number	Reaction	Reaction rate [cm^3 $molecule^{-1}$ s^{-1}]	References
1	$CO_2 + e^- \rightarrow CO + O\bullet + e^-$	$2*10^{-9}$	[6]
2	$CH_4 + e^- \rightarrow \bullet CH_3 + H\bullet + e^-$	$4.5*10^{-8}$	[5]
3	$CH_4 + e^- \rightarrow \bullet CH_2 + H_2 + e^-$	$7.3*10^{-9}$	[5]
4	$CH_4 + e^- \rightarrow \bullet CH + H\bullet + H_2 + e^-$	$3.7*10^{-9}$	[5]
5	$CH_4 + O\bullet \rightarrow \bullet CH_3 + \bullet OH$	$1.15*10^{-15}*T^{1.56}*e^{(-4275/T)}$	[7]
6	$\bullet CH_3 + O\bullet \rightarrow H_2CO + H\bullet$	$1.33*10^{-10}$	[7]
7	$H_2CO + O\bullet \rightarrow HCO\bullet + \bullet OH$	$3*10^{-11}*e^{(-1552/T)}$	[7]
8	$H_2CO + H\bullet \rightarrow HCO\bullet + H_2$	$3.65*10^{-16}*T^{1.77}*e^{(-1511.5/T)}$	[7]
9	$H_2CO + \bullet OH \rightarrow HCO\bullet + H_2O$	$1*10^{-11}$	[7]
10	$HCO\bullet + H\bullet \rightarrow CO + H_2$	$1.98*10^{-11}*T^{0.25}$	[7]
11	$HCO\bullet + \bullet O \rightarrow CO + \bullet OH$	$5*10^{-11}$	[7]
12	$HCO\bullet + \bullet OH \rightarrow CO + H_2O$	$1.67*10^{-10}$	[7]
13	$HCO\bullet + \bullet CH_3 \rightarrow CO + CH_4$	$8.3*10^{-11}$	[7]
14	$CO_2 + CO_2 \rightarrow CO + O + CO_2$	$4.39*10^{-7}*e^{(-128.6/RT)}$	[8]
15	$CO_2 + CO \rightarrow CO + O + CO$	$4.39*10^{-7}*e^{(-128.6/RT)}$	[8]
16	$CO_2 + O_2 \rightarrow CO + O + O_2$	$3.72*10^{-10}*e^{(-119.6/RT)}$	[8]
17*	$CO + O + CO_2 \rightarrow CO_2 + CO_2$	$6.54*10^{-36}*e^{(-4.3/RT)}$	[8]
18*	$CO + O + CO \rightarrow CO_2 + CO$	$6.54*10^{-36}*e^{(-4.3/RT)}$	[8]
19*	$CO + O + O_2 \rightarrow CO_2 + O_2$	$6.54*10^{-36}*e^{(-4.3/RT)}$	[8]
20	$O + CO_2 \rightarrow CO + O_2$	$7.77*10^{-12}*e^{(-33/RT)}$	[8]
21	$CO + O_2 \rightarrow CO_2 + O$	$1.23*10^{-12}*e^{(-25.2/RT)}$	[8]
22	$O_2 + O_2 \rightarrow O + O + O_2$	$8.14*10^{-9}*e^{(-118.6/RT)}$	[8]
23	$O_2 + O \rightarrow O + O + O$	$2.00*10^{-8}*e^{(-114.9/RT)}$	[8]
24	$O_2 + CO \rightarrow O + O + CO$	$2.40*10^{-9}*e^{(-118/RT)}$	[8]
25	$O_2 + CO_2 \rightarrow O + O + CO_2$	$2.57*10^{-9}*e^{(-111.5/RT)}$	[8]
26*	$O + O + O_2 \rightarrow O_2 + O_2$	$6.80*10^{-34}$	[8]
27*	$O + O + O \rightarrow O_2 + O$	$2.19*10^{-33}*e^{(4.5/RT)}$	[8]
28*	$O + O + CO \rightarrow O_2 + CO$	$2.75*10^{-34}$	[8]
29*	$O + O + CO_2 \rightarrow O_2 + CO_2$	$2.75*10^{-34}*e^{(-128.6/RT)}$	[8]

Running this kinetic mechanisms in Reactor Engineering Lab with a 1 mol/m^3 initial concentration for both CO_2 and CH_4 gives the results in figure 5, where all the concentration profiles are shown.

Journal of Physics: Conference Series **275** (2011) 012016 doi:10.1088/1742-6596/275/1/012016

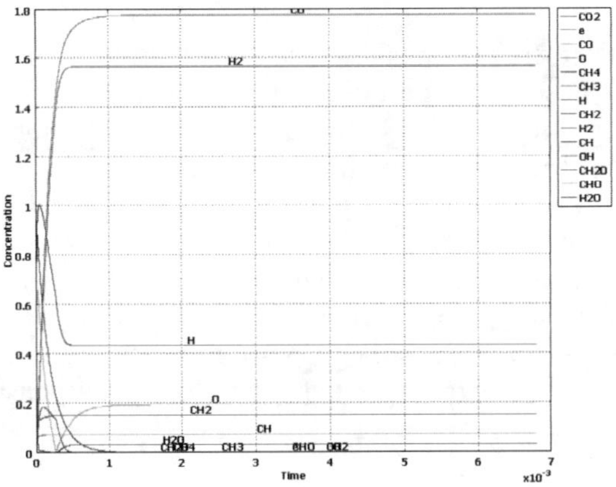

Figure 5: Concentrations as function of the time from the kinetic model of the CO_2 reforming of methane (table 1)

It can be seen that the main products are CO and H_2, followed by H and O radicals. This result suggests that it would be rather sufficient for the 3D reactor model to consider, next to the electrons, only the H and O radicals as active species in order to model the CO_2 reforming of methane.

3.2 The necessity of a 2D reactor model

Due to the large velocity field gradients, running the 3D model with the kinetic mechanism becomes not feasible. Two steps can be considered. The first step is to reduce the dimensional aspect of the modeling to 2 instead of 3. This is possible due to the axial symmetry of the reactor domain. Simulating the discharge region with a 2D reactor model allows to give a modified 3D reactor model (see later on) input values so that only the post-discharge model needs to be evaluated by the 3D reactor model. The kinetic mechanism in the discharge region concerns only the reactions and product emanating from the reaction between CO_2 and the electrons. Therefore, these reactions are taken from table 1 (reactions 1 and 14 to 29). The second step is using a strongly reduced mechanism for the reformation of methane by active species coming from the discharge region. The first step is considered in this section and the step in section 4. Figure 6 presents the mesh of the 2D reactor model (which is a "slice" of the 3D reactor model from figure 2).

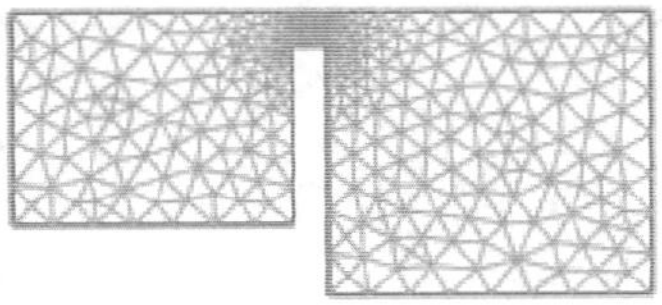

Figure 6: The mesh of the 2D reactor model

Note that in figure 6, the mesh has been turned 90 degrees. So the left vertical line represents the inlet (where the CO_2 enters), though a bit differently. The middle (small part) is where the electron source is introduced. The right part represents the post-discharge region. The same type of "ns" and "chdh" models are used for respectively the Navier-Stokes and convection-diffusion equations. As said before, the inlet boundary limit in the discharge region should be 0.021 m/s. Since the inlet is not exactly the same, the velocity has been changed (it appeared to be 0.0027 m/s), in order to obtain the value of 0.021 m/s at the inlet of the discharge region.

3.3 Reduction of the kinetic mechanism

In order to render the 2D calculation more feasible with sufficient precision, the kinetic mechanism for the discharge region (reactions 1 and 14 to 29 in table 1) is somewhat reduced by eliminating the reactions that have relatively low reaction rates in combination with relatively low product concentrations. The remaining reactions are given by reactions 1, 14 and 15 of table1, containing 3 reactions and 4 species. The CO and O• profiles calculated by this kinetic mechanism showed good quantitative agreement, so that it is not necessary to present the figures. These 3 reactions are then introduced into the 2D reactor model, of which the results are presented in the next section. It should be noted, though, that a too strong reduction of a kinetic mechanism can lead to less precise representation of the chemistry. However, the necessary reduction in order to include kinetics into a CFD code, still gives more reliable and correct results than a more detailed mechanism where aerodynamics are neglected. Therefore, it is considered acceptable using such a reduced mechanism for the purposes of this paper.

3.4 Concentrations of the active species in the discharge region

Before any modeling can be made, it is important to calculate the electron density in the discharge region, since this influences considerably the kinetics, the electrons being energetic initiators of the chemical process. A 1D Matlab program (developed in the laboratory LGPPTS) serves for calculating the concentration of electrons and ions. It is mainly based on solving non-linear equations of conservation for electrons and ions and the Poisson equation for the electric potential. This model predicts an electron density of 10^{14} cm^{-3}. This electron density is implemented in the 2D reactor model by entering in the subdomain a constant source term for the electron production. In the left domain, an initial CO_2 concentration of 1 mol/m^3 is maintained. Figure 7 shows the velocity field and the concentrations of the electrons, the H radicals and the O radicals.

Figure 7: From left to right: the velocity field (0 – 0.0236 m/s), the electron concentration (0 – $2.297*10^{-3}$ mol/m^3), the O radicals concentrations (0 – 1 mol/m^3) and the CO concentration (0 – 1 mol/m^3)

For the 3D reactor model, the values for the input at the discharge region are taken to be 0.0236 m/s, $2.297*10^{-3}$ mol/m^3, 1 mol/m^3 and 1 mol/m^3 for respectively the velocity field, the electron concentration and the O and CO radicals concentration.

4. The CO_2 reforming of methane

4.1 Reduction of model

The same reduction procedure is done for the methane part of table 1 (reactions 2 to 13). This gives a kinetic mechanism, described by reactions 2, 3, 6, 8, 10 and 13 of table 1, containing 6 reactions and 9 species. Figure 8 shows the results of running this mechanism together with the one in section 3.3 in Reaction Engineering Lab. Good agreement can be found between the reduced model and the initial model.

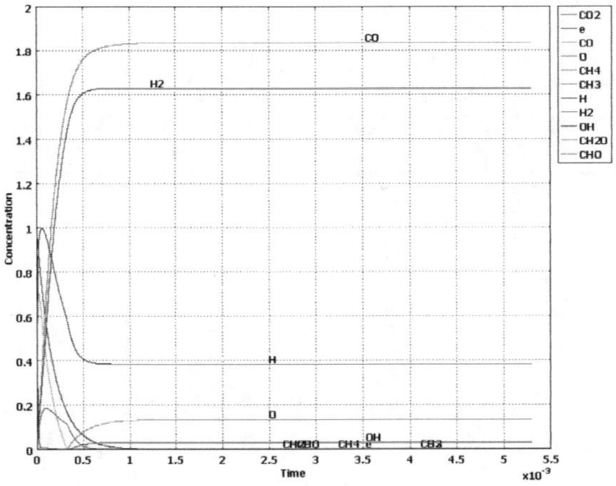

Figure 8: Concentrations as function of the time from the kinetic model of the CO_2 reforming of methane (table 2)

4.2 The 3D modeling of the post-discharge region

The 3D model in question corresponds to the upper half of the reactor depicted in figure 2 above the discharge region. The values found previously in the discharge region give therefore the inlet boundary limits for the velocity (0.021 m/s) and the concentrations of the electrons and CO:O radicals (respectively $2.297*10^{-3}$ and 1:1 mol/m^3). The boundaries, subdomains and mesh of this reactor are shown in figure 9. The velocity field is also represented using the same "ns" and "chdh" models as before. This model is then run with the inlet conditions presented above (note that the inlet velocity of the CH_4 was set to 1 m/s).

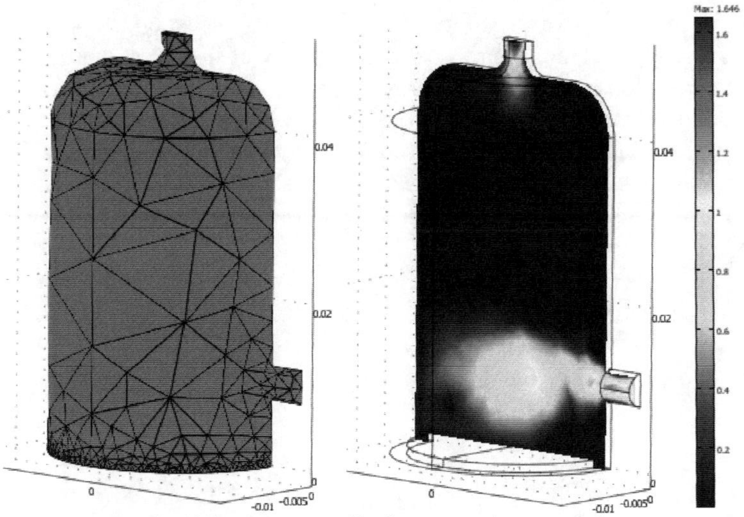

Figure 9: Mesh of the 3D post-discharge region model

4.3 Results of the 2D reactor model

The reduction of the mechanism in the previous section is not enough for implementation into the 3D reactor model. Therefore, the model had to be reduced further by eliminating the reactions that contain constituents with lower concentrations where one reaction is faster than the subsequent, the former having a constituent as a product, which is the source for the latter (Quasy-Steady-State-Assumption). Then the remaining reactions that follow one pathway ($CH_4 \rightarrow CH_3 \rightarrow CH_2O \rightarrow CHO \rightarrow CO + H_2$) are regrouped to one overall reaction leading from CH_4 directly to CO and H_2, taking the reaction rate of the initiation reaction (reaction 2). The reduced mechanism, consisting now of 2 reactions and 6 species (electrons included), is shown in table 2. Figure 10 shows the concentrations of the remaining species.

Table 2: Reduced mechanism.

1	$CO_2 + e^- \rightarrow CO + O\bullet + e^-$	$2*10^{-9}$
2+6+8+10	$CH_4 + O\bullet + e^- \rightarrow CO + 2H_2 + e^-$	$4.5*10^{-8}$

As figure 10 shows, the quantitative values do not agree well. This is just the problem with implementing kinetic mechanisms in 3D codes. However, qualitatively, the values are acceptable and the order of magnitude does not differ much with that of Figs. 5 and 7. Furthermore, such a drastic (loosing precision) reduction is necessary in order to use this kinetic model into the 3D reactor model, of which the results are presented in figure 11. The purpose of this 3D reactor model is, therefore, not to give precise values, but rather investigate the global characteristics and potentials of CO_2 reforming of methane in such a reactor.

First, the values of the CH_4 concentrations were, except for the boundary entrance limit, not significant. Therefore, its results are not presented. It can be seen that hydrogen and CO are mainly formed at the level of the entrance, though more hydrogen is formed (note that the CO concentration was already 1 mol/m^3 when leaving the discharge region). It seems that the reactor aerodynamics favor the formation of hydrogen over CO. The concentration of the O radical is also at its highest around the entrance of the reactor, when not considering the O radical concentration at the exit of the discharge region. Moreover, due to the latter situation, it appears that the peak concentration (disregarding the discharge region exit) of the O radical is shifted from the entrance level towards the reactor wall. This corresponds also with the peak concentration of CO. Hydrogen does not seem to be present significantly at that spot, since it comes only from methane, resulting into the formation of hydrogen at the entrance level. CO, however, is formed at both the entrance level and towards the wall. These results represent well (qualitatively) the kinetic mechanisms, having hydrogen formed from the interaction between the O radical, electrons and methane, while CO also has a source near the reactor wall (discharge region exit).

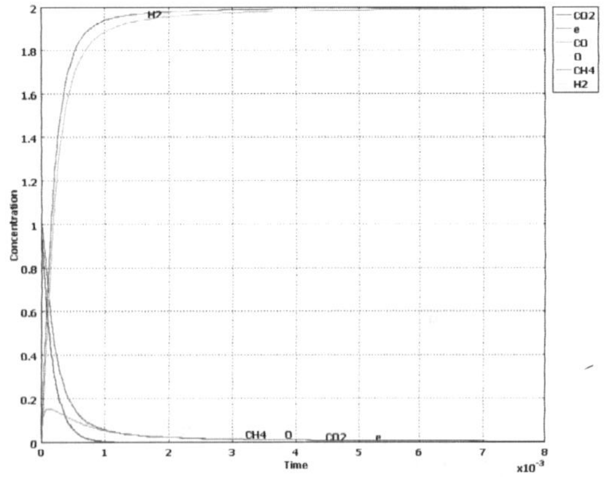

Figure 10: Concentrations as function of the time from the reduced kinetic model of the CO_2 reforming of methane (table 3)

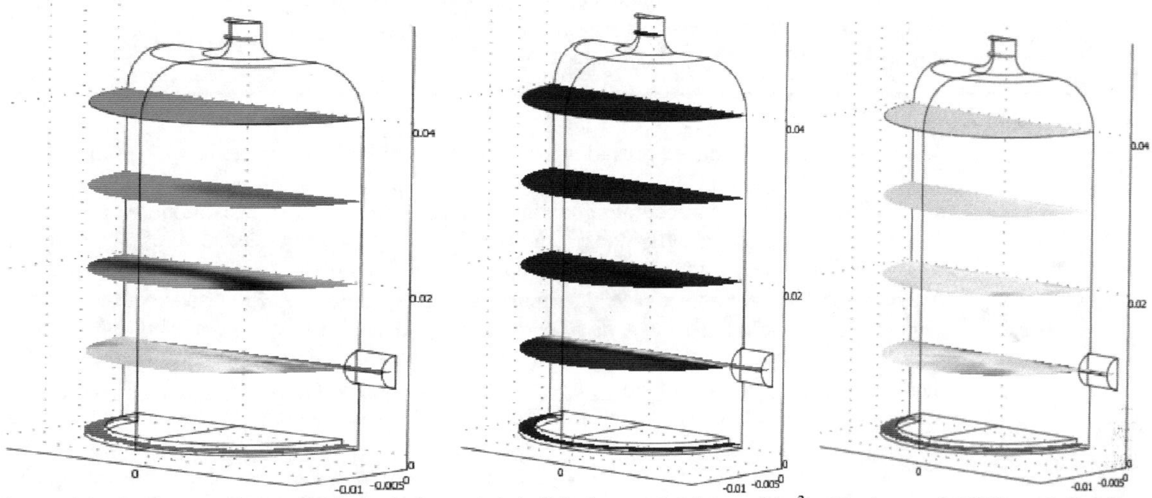

Figure 11: Concentrations of (from left to right) CO (max. 1.296 mol/m^3), H_2 (max. 2.592) and O• (max. 1.244 mol/m^3), using the 3D post-discharge reactor model. The red color shows the maximum concentrations and the blue color stands for 0 mol/m^3.

5. Conclusions

The CO_2 reformation of methane in a Dielectric Barrier Discharge reactor can be modeled quite satisfactorily in a 3D model, provided that the kinetic model is strongly reduced to an extent that the precision decreases. For more kinetic precision, a 2D model offers acceptable results. The aerodynamic flow, convection-diffusion characteristics are shown in the discharge region. For more successful 3D modeling, a higher computer performance is needed.

References

[1] Malik A, Jiang X Z 1999 The CO_2 Reforming of Natural Gas in a Pulsed Corona Discharge Reactor, *Plasma Chem. Plasma Proc.*, **19** 4.

[2] Kraus M, Egli W, Haffner K, Eliasson B, Kogelschatz U, Wokauna A 2002 Investigation of mechanistic aspects of the catalytic CO_2 reforming of methane in a dielectric-barrier discharge using optical emission spectroscopy and kinetic modeling, *Phys. Chem. Chem. Phys.*, **4** 668–675.

[3] Eliasson B, Kogelschatz U, Xue B, Zhou L M 1998 Hydrogenation of Carbon Dioxide to Methanol with a Discharge-Activated Catalyst, *Ind. Eng. Chem. Res.* **37** 3350-57.

[4] Oumghar A, Legrand J C, Diamy A M, Turillon N 1995 Methane Conversion by an Air Microwave Plasma, *Plasma Chem. Plasma Proc.*, **15** 1.

[5] Kline L E, Partlow W D, Bies W E 1989 Electron and chemical kinetics in methane rf glow-discharge deposition plasmas, *J. Appl. Phys*, **65** (1).

[6] Eliasson B, Egli W, Kogelschatz U 1994 Modelling of dielectric barrier discharge chemistry, *Pure & Appl. Chem.*, **66** 1275-86.

[7] Marinov N M, // Int. J. Chem. Kinet. 1999. Vol.31. P.183-220.

[8] Fridman A, Plasma Chemistry, Cambridge University Press, 2008.

Comparison of LIBS and μ-XRF measurements on bronze alloys for monitoring plasma effects

M F Alberghina, R Barraco, M Brai, T Schillaci* and L Tranchina
Dipartimento di Fisica e Tecnologie Relative,Università di Palermo, Viale delle Scienze Ed. 18, 90128 Palermo (Italy)

Abstract. The laser-induced breakdown spectroscopy (LIBS) technique is often used as atomic spectroscopic technique for elemental analysis of materials. However, it presents some drawbacks that make an accurate quantitative analysis difficult. Since the plasma properties, such as spatial inhomogeneity and plume stoichiometry strongly depend on the experimental conditions, the measurements are less reproducible. In order to evaluate the measurement fluctuations, we propose to use the more established micro X-Ray fluorescence (μ-XRF) technique for validating LIBS data. In particular, the quantitative data, obtained by varying the laser fluence, the shot numbers and the temporal acquisition parameters, were compared with those obtained by μ-XRF on laboratory made samples of binary, ternary and quaternary bronze alloys. For LIBS measurements a mobile double pulse laser instrument equipped with an high resolution Echelle type monochromator coupled to an intensified CCD camera was used. μ-XRF analyses were performed with a portable instrument that uses a micro collimated X-Ray beam and it is equipped with an high resolution detector. The LIBS results show a strong dependence both on the instrumental set up and the chemical-physical properties of the sample. With our findings we could identify the most suitable parameters to be used in the investigation of the different bronze alloys. The possibility to carry out a quantitative analysis by using the LIBS technique was checked through the comparison with related μ-XRF data. In particular in this paper we identified a set of reliable LIBS parameters for the quantitative analysis of copper, tin and zinc. Further analyses will be necessary to reach this goal also for the minor constituents as lead.

Keywords: Plasma features, LIBS parameters, μ-XRF, bronze alloys, quantitative analysis.

1. Introduction

Laser-induced breakdown spectroscopy (LIBS) is a spectroscopic technique based on laser-induced formation of a plasma above the sample surface, followed by time-resolved analysis of the optical emission associated to atomic and molecular transitions. In the past two decades, LIBS has developed rapidly as an analytical approach to elemental analysis, especially in process-analytical applications, microanalysis and remote sensing [1-3]. The reliability and refinement of LIBS as an analytical tool was made possible by a continuous improvement in instrumentation, such as the intensified charge couple device (ICCD) array detector and more mature and more reliable laser sources. The array detectors are very important since they allow the capture of multiple emission lines from a single LIBS event. The advancement in the development of broadband high resolution spectrometers, including Echelle and multi-spectrometers design made the detection and the simultaneous analysis of multiple key elements of the target material possible. Up to now this technique has mainly been employed for

* Corresponding author.
E-mail address: tschillaci@unipa.it
Tel.: +39 091 23899149; Fax.: +39 091 6615069

the qualitative analysis. The quantitative applications are difficulty due to several factors that affect the measurements leading to the instability of the laser-induced plasma. Main parameters affecting he analytical performance of LIBS are the laser properties (wavelength, energy, pulse duration, shot-to-shot power fluctuation), the ambient conditions, the thermo-physical properties of the sample, the detection window (delay time and gate width) and the geometrical set-up of the collecting optics [4]. These problems have renewed the need for qualitatively and quantitatively understanding the laser-interaction process in order to determine the optimal experimental conditions for analytical measurements [5]. The initiation, formation and decay of the plasma are complex mechanisms. As the laser plasma is the excitation source of LIBS measurements, the properties of the laser can influence the analytical qualities of the LIBS analysis. After the laser pulse has terminated, the plasma decay over an interval from few hundred nanoseconds up to several microseconds, depending on the laser energy deposited and on the material composition [6]. At higher laser fluences, different mechanisms come into play because, for subsequent pulses, the laser energy hits the melted and recondensed matter. The elements of the original sample may have been selectively evaporated so that the redeposited material no longer exhibit the original composition [7]. The spectra observed change as the plasma evolves temporally and the goal of retaining the composition of the sample after ablation into the plasma is an important issue. Moreover, for an accurate chemical analysis, the ablated mass composition must be the same as the sample composition [6, 7].

When the plasma is in local thermal equilibrium (LTE), local gradient of plasma properties (temperature, density, thermal conductivity) is low enough to let a particle in the plasma to reach the equilibrium: diffusion time must be similar or higher than the time necessary to the particle to achieve the equilibrium [8]. LTE of the plasma is a condition that maximizes the analytical ability of the plasma as source of a quantitative physical technique. On the basis of the above considerations, an investigation approach to made the experimental conditions similar to the LTE is to set the LIBS parameters in a way to reduce the statistical fluctuations and to have an experimental mean value of the relative intensity ratio of two spectral lines, arising from the same upper level, closest to the theoretical value. In this way a quantitative analysis could be feasible. On the other hand, we have to consider also the temporal evolution of the plasma that is related to the thermo-physical properties of the element present in the sample. In particular, due to the boiling temperature and to the heat of vaporization, the pressure vapor of each element in the sample could assume different values and different decay times. This makes the stoichiometric composition of the plasma different from that of the original sample [7]. Then, in order to optimize the LIBS measurement procedure it is necessary to verify that the LTE condition is satisfied and to evaluate the detection delay time for each element to be investigated. For this purpose we studied copper-base alloys, difficult to analyse by LIBS because of their large dissimilarity in the chemical-physical properties of the elemental constituents which strongly affect the stoichiometric ratios with respect to the original evaporated surface [7-10]. Nevertheless, the LIBS technique, applied as quantitatively analytical analysis needs to be checked from a more routinary and well established technique such as the XRF [11-13]. Then, the LIBS data were compared with those obtained by XRF in order to verify the reliability of the quantitative LIBS analysis. On the basis of our findings, we will propose to extend this methodological approach to other kinds of materials such as stones, glasses, biological samples and so on.

2. Experimental

Five laboratory made metal samples containing binary, ternary and quaternary alloys were analyzed. The nominal composition of the samples is reported in Tab. 1 as weight percentage. The samples were mechanically lapped on a surface by a swing-frame grinder and chemically finished by using abrasive paste in order to obtain a mirror surface. This was subsequently polished by repetitive ultrasound washing processes.

Table 1. Nominal composition of the bronze samples.

Sample label	Cu (%)	Sn (%)	Pb (%)	Zn (%)
DFT 1	85.0	15.0	–	–
DFT 2	95.0	5.0	–	–
DFT 3	97.0	3.0	–	–
DFT 4	92.3	7.5	0.2	–
DFT 5	82.3	3.0	0.5	14.0

LIBS analysis was carried out with a portable double-pulse instrument (Marwan Technologies, mod. Modì). Laser ablation of the samples was obtained by means of a double Q-Switched Nd:YAG laser (Lotis TII, mod. 2131 DM). LIBS measurements were performed on the samples located inside a closed experimental chamber, equipped with a motorized table for the exact positioning of the sample into the focus of the two laser beams used for sample positioning. An optical microscope allowed to monitor the region of the sample under analysis.

The LIBS instrument integrates a Nd:YAG dual-pulse laser that emits two collinear laser pulses at 1064 nm with a maximum repetition rate of 10 Hz. The reciprocal delay is adjustable from 0 to 80 μs with a laser duration of about 10 ns. In our investigation, the flash lamp energy was set in the range 9.5 ÷ 23 J with an output laser energy per pulse of 5 ÷ 120 mJ. The laser beam was focused on the sample by using a Plano-convex lens. Plasma emission is collected to an angle of about 30° respect to the laser beam axis and carried by an Y bundle optical-fiber. One fiber end is linked to the entrance slit of a compact Mechelle spectrometer (Andor, mod. 5000), whereas the other end to a gated ICCD (Andor, mod. iStar). The gate aperture of the intensified CCD was synchronized with the laser burst. Mechelle spectrometer, coupled to the gated ICCD, allows the simultaneous spectral recording in the range 200 ÷ 950 nm with a resolving power ($\lambda/\Delta\lambda$) of 5000. LIBS instrument is controlled via an integrated personal computer which manages sample visualization and positioning, the experimental settings of the laser - energy of the beams, delay between the pulses, repetition rate - and the spectral acquisition parameters - number of spectra averaged, gate pulse delay (GPD), gate pulse width (GPW) and multi channel plate (MCP) gain. The experimental set-up values used for the laser source and for the Mechelle spectrometer were chosen with the aim to enhance the signal-to-noise ratio (SNR). In order to evaluate the dependence of crater features (diameter, morphology and fusion state of the metal) on laser energy, LIBS acquisitions were performed at different laser energy values in the range 5÷120 mJ per pulse, with one or five double laser shot accumulations, setting the ICCD setup gater GPD and GPW to 500 ns and 2.0 μs respectively. SEM images on the craters were acquired in high vacuum modality by using a Quanta 200 F (FEI Company) microscope. The set up instrument were 15 kV for the electron gun, 300× magnification, and a chamber pressure of 7.87 10^{-5} mbar. Crater images were also acquired by optical microscopy (OPTEK) with a magnification of 50×. Moreover, LIBS acquisitions were performed by setting the laser energy value at 15, 30 and 70 mJ, the ICCD setup gater GPD variable in the range 0.5÷10 μs and GPW in the range 1.5÷3.5 μs. All the measurement were performed with a delay time between two laser pulses of 1 μs and setting the delay time of the first Q-Switch laser pulse to 135 μs. The MCP gain of the ICCD was setting to 230. A calibration of the Mechelle Spectrometer was performed before each set of measurement by using an Hg-Ar calibration lamp (Ocean Optics).

LIBS spectra, after acquisition and storage, are qualitatively and quantitatively analyzed using a dedicated software (LIBS++). The calibration-free method [14], implemented by LIBS++, was not used because it is based on the assumption that the plasma is in LTE, condition that instead we intend to verify. Thus, measurements of the peak intensity areas were performed for each spectral line characterizing the elements from the LIBS spectra. The peak emission lines used were 510,5 nm,

515.3 nm and 521.8 nm for Cu (I), 468.0 nm, 472.2 nm and 481.0 nm for Zn(I), 364.0 and 368.3 nm for Pb(I), 283.9 nm and 286.3 nm for Sn(I). The more intense peak emission lines have been considered for the evaluation of the transition probability ratios. In particular, 515.3 nm and 521.8 nm for copper, 283.9 nm and 286.3 nm for tin and 472.2 nm and 481.0 nm for zinc.
XRF analyses were performed by using a portable X-ray fluorescence (XRF) instrument ArtTAX 200 (Bruker AXS). The ArtTAX is equipped with a low-power X-Ray tube with a molybdenum anode. A pinhole system, in the X-ray source, provides collimated beam in the sample making available the instrument to perform spatially resolved multi-elemental analysis on three-dimensional structures. The X-ray fluorescence is detected by a XFlash detector (silicon drift detector) with high speed and low noise electronics with an energy resolution <145 eV at 5.9 keV. ArtTAX is equipped with a helium flow system that allows to detect the lighter elements up to sodium. The helium flow on the detection path reduces the photoelectric absorption of characteristic X-rays emitted by the sample due to air molecules. The experimental XRF settings was chosen in order to optimize the SNR and we used: for the molybdenum anode voltage and current, 40 kV and 600 μA; an acquisition time of 300 sec and a helium flow rate of 1.6 l/min.

3. Results and discussions

In order to evaluate the dependence of crater features (diameter, morphology and fusion state of the metal) we present in Fig. 2, the SEM images of the sample craters related to four values of the flash lamp energy (9.5 J, 11.5 J, 13.5 J and 23 J). The comparison of the SEM images shows that the sample DFT 1 presents a more regular circular shape and a complete melting process of the material. This fact is due to the thermo-physical properties and relative concentrations of copper and tin present in DFT 1 alloy. Table 2 reports the thermo-physical properties of all elements contained in the samples. In particular, the lowest melting heat of tin and the highest thermal conductivity of copper are the main factors that affect the crater features. The first one causes a decreasing of alloy melting temperature, the second one a faster heat loss, both factors lead to a more homogeneous fusion state and shape.

Table 2. Thermo-physical properties of the elements present in samples.

	Pb	**Sn**	**Zn**	**Cu**
Atomic mass	207.2	118.71	65.39	63.546
Thermal conductivity : ($Wm^{-1} K^{-1}$)	35.3	66.6	116	401
Melting temperature (k):	600.5	504.9	692.5	1357.6
Boiling temperature (k):	2032	2865	1180	2835
Heat of melting: ($KJ\ mol^{-1}$)	5.121	7.2	6.67	13
Heat of vaporization: ($KJ\ mol^{-1}$)	177.8	296.2	114.2	306.7
Specific heat capacity: (J/kg/K)	130	210(α) 220(β)	390	386

On the basis of the optical images, the area evaluation of the laser crater for DFT 1 sample, shown in Fig. 3 as function of the laser energy has been obtained using ImageJ, a Java-based image processing software [15]. The plot evidences the linear trend of the areas with increasing laser energy. Moreover this tendency is noticeable also in other samples, even if due to the above reasons, it is affected by larger statistical error. These facts can be correlated with the peculiar thermal properties of the different compositional matrix of the samples. In fact, samples DFT 1, 2 and 3 are binary Cu-Sn alloys, DFT 4 and DFT 5 are ternary and quaternary alloys, respectively.
A typical LIBS spectrum, is showed in Fig. 4, that displays two wavelength ranges of the same LIBS acquisition on DFT 5 sample. In this figure are underlined the spectral lines considered in this work. In

order to highlight the lower spectral line intensities corresponding to Sn (I) and Pb(I) respect to Zn(I) and Cu(I) which are ten times more intense, the intensity scales on the figures were settled to different values. Focusing our attention on DFT 1 sample, the behaviour of the peak intensity values of copper neutral atom Cu (I) emission lines at 515.3 nm and 521.8 nm and tin neutral atom Sn (I) emission lines at 283.9 nm and 286.3 nm as function of the laser flash lamp energy values are shown in Fig. 5 respectively.

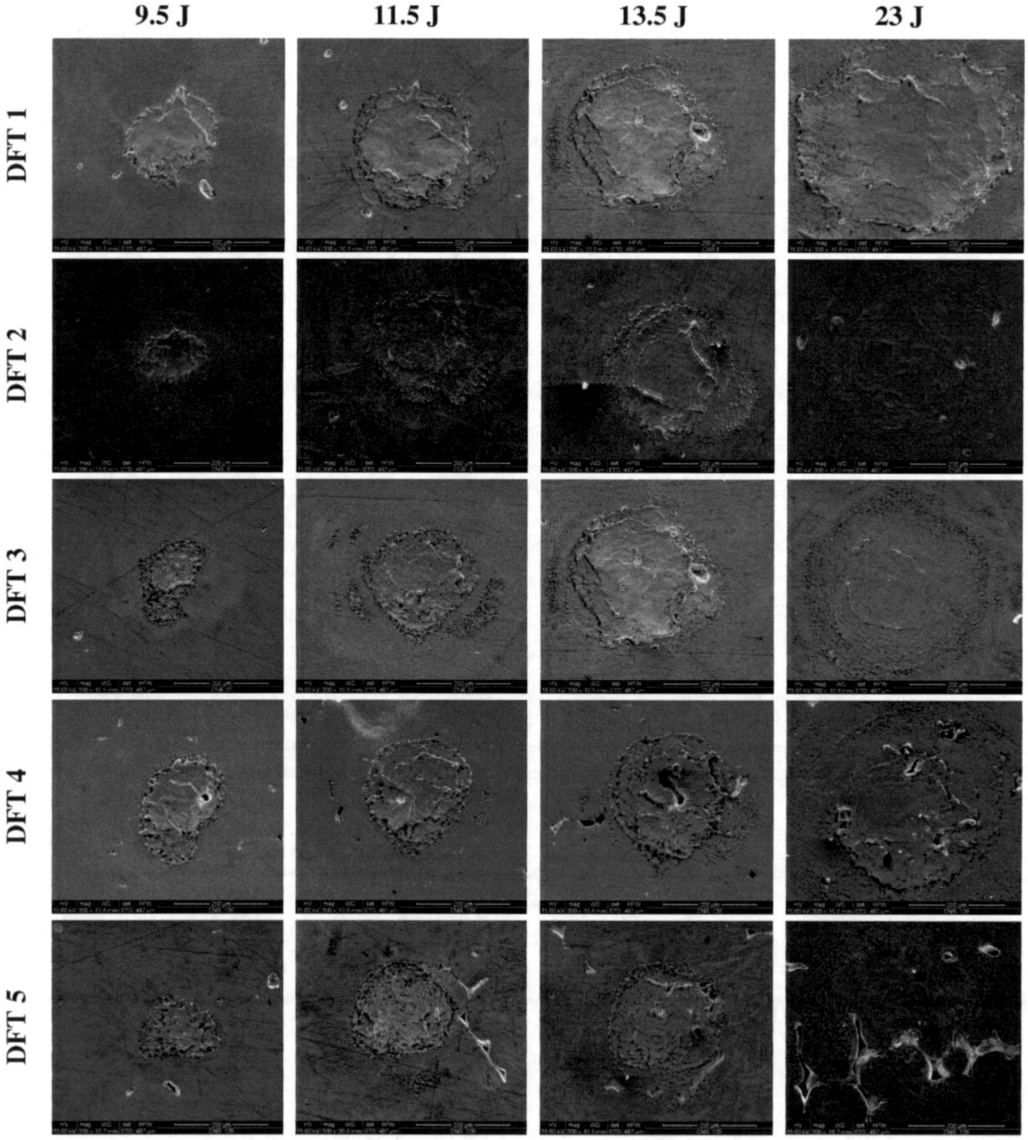

Figure 2. SEM images of sample craters related to four laser flash lamp energy values

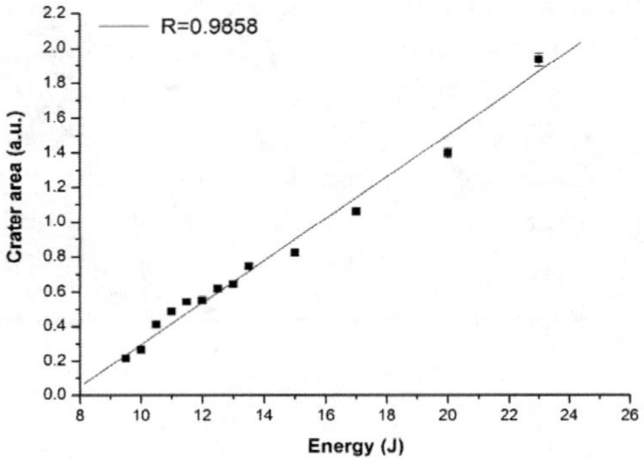

Figure 3. The crater area as a function of laser flash lamp energy.

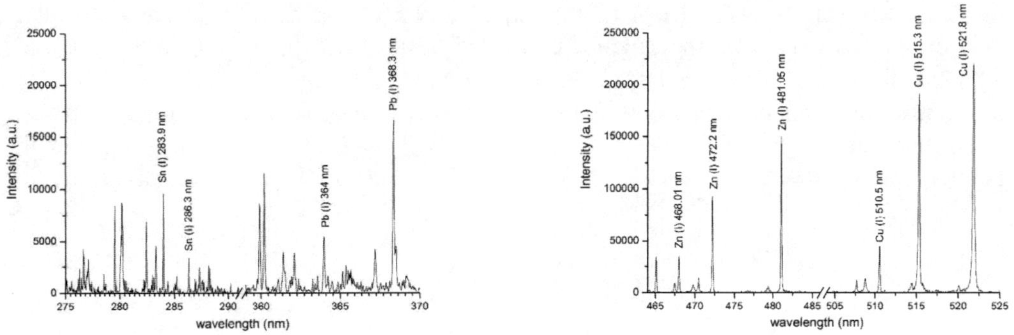

Figure 4. LIBS spectrum of DFT 5 sample. On the left the wavelength region with evidenced the Sn (I) and Pb(I) spectral lines and on the right the spectral range for Zn(I) and Cu(I).

The choice of the suitable laser energy is conditioned by the need to have the best SNR and to minimize the LIBS micro destructive aspects, mandatory for particular applications where it is necessary to guarantee the integrity of objects to be investigated. The flash lamp energy values that satisfy the above requirements are in the rage of 10.5 ÷ 12.5 J within we selected out the mean value of 11.5 J, evidenced as red in Fig. 5.

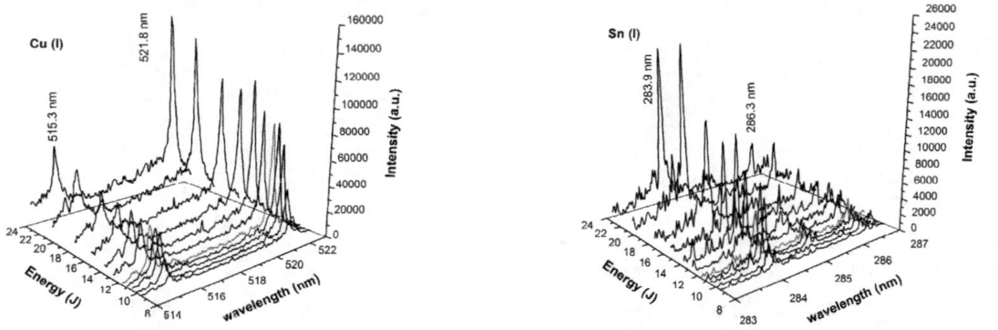

Figure 5. Line intensities as a function of the laser flash lamp energy for Cu (I) (spectral lines 510.5 nm, 515.3 nm and 521.8 nm) and Sn (I) (spectral lines 283.9 nm and 286.3 nm).

Journal of Physics: Conference Series **275** (2011) 012017 doi:10.1088/1742-6596/275/1/012017

In order to optimize the temporal ICCD parameters, GPW and GPD, different measurements were carried out with the chosen 11.5 J laser flash lamp energy value. The optimal GPW value of the Mechelle spectrometer for detecting copper in the samples, relatively to transition probabilities of Cu ($3d^{10}\ ^1S\ 4d\ ^2D_{3/2} \rightarrow 3d^{10}\ ^1S\ 4p\ ^2P_{1/2}$ and $3d^{10}\ ^1S\ 4d\ ^2D_{5/2} \rightarrow 3d^{10}\ ^1S\ 4p\ ^2P_{3/2}$, corresponding to Cu(I) emission lines at 515.3 nm and 521.8 nm), has been determined, using the relative line intensity ratio of two spectral lines arising from the same upper level. The relative intensity ratio can be calculated by [16]

$$\left(\frac{I_1}{I_2}\right)_{th} = \frac{A_1 g_1 \lambda_2}{A_2 g_2 \lambda_1} \tag{1}$$

where $A_{1,2}$ indicates the transition probability between the state two state, $g_{1,2}$ the statistical weight and $\lambda_{1,2}$ the corresponding wavelength. The theoretical relative intensity ratio, 1.75 for Cu [17,18], has been compared with the average value of the experimentally obtained transition probability ratios. These are reported in table 3 for the samples that have the lowest copper concentration as the DFT 1 and the DFT 5. In table 3, σ corresponds to the standard deviation.

The variation of the transition probability ratios were calculated at different delay times. Table 3 shows also the relative deviation Δ defined by equation (2). Δ measures the adherence of the mean value to the theoretical one.

$$\Delta = 1 - \frac{(I_1 / I_2)_{meas}}{(I_1 / I_2)_{th}} \tag{2}$$

The value of Δ, optimal for both samples, allowed us to select 2.5 μs as the more suitable value of GPW for copper.

Table 3. Transition probability ratios of Cu (I) lines.

	DFT 1		**DFT 5**	
GPW [μs]	$(\mathbf{I_1/I_2})_{\mathbf{meas.}} \pm \sigma$	**Δ**	$(\mathbf{I_1/I_2})_{\mathbf{meas.}} \pm \sigma$	**Δ**
1.5	1.96 ± 0.09	- 0.12	2.13 ± 0.23	- 0.21
2.0	1.97 ± 0.15	- 0.13	2.29 ± 0.20	- 0.30
2.5	2.04 ± 0.17	- 0.16	1.59 ± 0.26	0.09
3.0	2.52 ± 0.31	- 0.44	1.69 ± 0.13	0.04
3.5	2.79 ± 0.49	- 0.59	1.53 ± 0.19	0.13

In order to identify the optimal value of GPW of the Mechelle spectrometer to detect the tin (present only in the DFT 5 sample), the relative transition probabilities of Zn triplet ($4s5s\ ^3S_1 \rightarrow 4s4p\ ^3P_{0,1,2}$) has been determined, using the same above criteria. The theoretical relative intensity ratios, 1.50, 4.39 and 2.93 for Zn [18], have been compared with the average value of the experimentally obtained transition probability ratios. The results are given in table 4 for the DFT 5 sample. The value of Δ allowed us to select 2.5 μs as more suitable value of GPW for Zn, in accordance also with the possibility to reveal with the same ICCD parameters copper and zinc present in the bronze alloys.

Table 4. Transition probability ratios of Zn (I) lines.

	1.50		**4.39**		**2.93**	
GPW [μs]	$(I_1/I_2)_{meas.} \pm \sigma$	Δ	$(I_1/I_2)_{meas.} \pm \sigma$	Δ	$(I_1/I_2)_{meas.} \pm \sigma$	Δ
1.5	2.05 ± 0.54	- 0.36	4.35 ± 0.64	0.01	2.18 ± 0.35	0.26
2.0	1.92 ± 0.19	- 0.28	4.59 ± 0.67	- 0.04	2.39 ± 0.26	0.18
2.5	1.69 ± 0.18	- 0.12	4.30 ± 0.71	0.02	2.57 ± 0.44	0.12
3.0	1.71 ± 0.25	- 0.14	3.83 ± 0.77	0.13	2.34 ± 0.37	0.23
3.5	1.65 ± 0.11	- 0.10	4.54 ± 1.07	- 0.03	2.75 ± 0.56	0.06

The variations of the spectral line intensities of Cu doublet ($3d^{10}\,{}^1S\,4d\,{}^2D_{3/2} \rightarrow 3d^{10}\,{}^1S\,4p\,{}^2P_{1/2}$ and $3d^{10}\,{}^1S\,4d\,{}^2D_{5/2} \rightarrow 3d^{10}\,{}^1S\,4p\,{}^2P_{3/2}$) and Zn triplet ($4s5s\,{}^3S_1 \rightarrow 4s4p\,{}^3P_{0,1,2}$) with the GPD times are shown in Fig. 6 relatively to the DFT 5 sample.

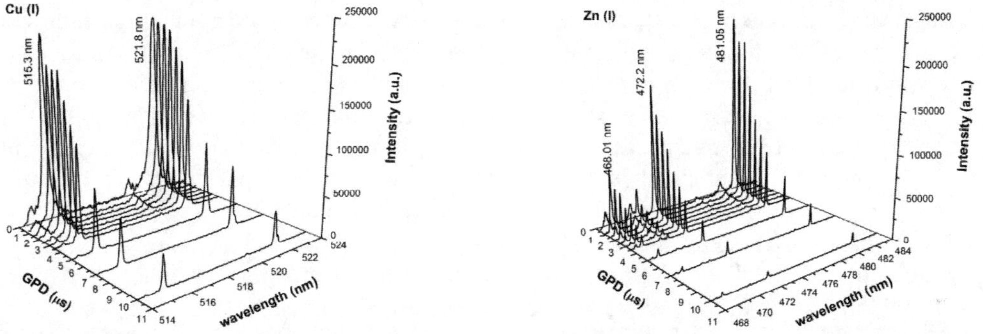

Figure 6. Variations of the line intensities of Cu (I) and Zn(I) as a function of the GPD times.

The average value of the experimental transition probability ratios for the copper doublet and the zinc triplet are given in table 4, whereas its variation with the delay time is shown if Fig. 7.

Table 4. Transition probability ratios of Cu (I) and Zn (I) lines.

Element	**Wavelength (nm)**	**Experimental**	**Theoretical (NIST [18])**
Cu (I)	521.8/515.3	1.59 ± 0.26	1.75 ± 0.44
	481.05/472.2	1.69 ± 0.18	1.50 ± 0.38
Zn (I)	481.05/468.01	4.30 ± 0.71	4.39 ± 1.10
	472.2/468.01	2.57 ± 0.44	2.93 ± 0.73

For sake of clarity, in Fig. 7 are reported only the transition probability ratios of the zinc spectral lines 472.20 nm on 468.01 nm and 481.05 on 472.2 nm. GPW value was settled to 2.5 μs.

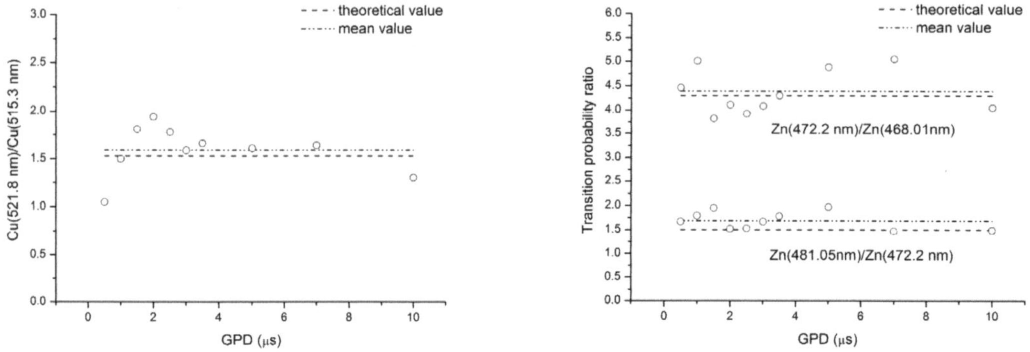

Figure 7. Transition probability ratios in the copper plasma decay (on the left) and in the zinc plasma decay (on the right).

Evaluation of tin and lead was not possible with the chosen set up parameters. In fact, different partial vapor pressure values as a function of the time delay of the elements present in alloy, may lead to the plasma non-stoichiometry. In particular, the partial vapor pressure of tin and lead, which have a lower boiling temperature than copper [7], reaches a value less than two order of magnitude respect to copper. The temporal evolution of both elements has a similar behaviour with a decay time of the order of few tens ns. Therefore, for revealing lead and tin, it is necessary to reduce the GPD time in order to detect the largest amount of the signal emitted from the plasma, but on the other hand, this reduction on GPD value leads to a decrease in the SNR. To overcome this problem, a compromise was reached by carrying out measurements with a greater number of LIBS acquisitions and increasing the laser energy value. Consequently, five laser shots were accumulated, setting the laser flash lamp energy to16.5 J, the GPD to 1.5μs and the GPW to 2.5 μs.

In Fig. 8 are reported the LIBS calibration curves for tin and copper. The plots report the peak intensity ratio as a function of tin and copper weight concentration relatively. In Fig. 8 on the right, the first value of the calibration curve of copper and related to the DFT 5 sample, suffer from the fact that the partial pressure vapour of zinc has a decay time faster than the other elements present in the alloy and of the order of five times greater. This means that the concentration of zinc detected in the plasma is over estimated in comparison to the real concentration on the sample.

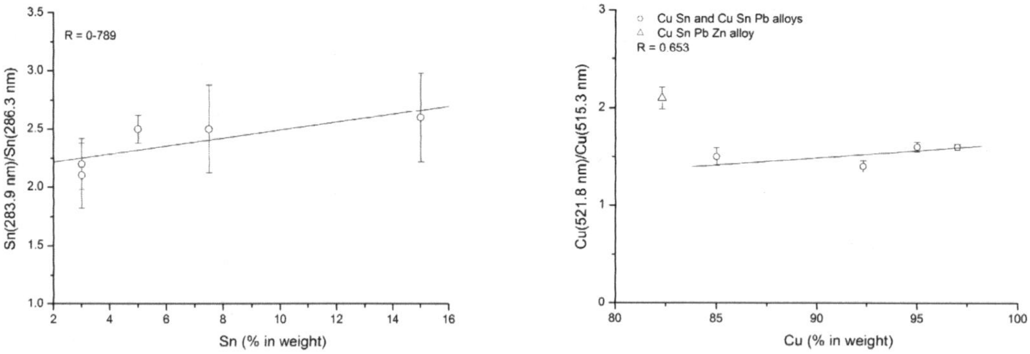

Figure 8. LIBS calibration curves reported as a function of tin weight percentage (on the left) and copper weight percentage (on the right).

Fig. 9 shows the analytical copper peak intensity normalized to tin peak intensity, as a function of the element concentration ratio in the samples. Experimental data were well fitted by a linear equation relatively to binary and ternary bronze alloys, being the data of the quaternary alloy, as before mentioned, influenced from the non stoichiometric composition of the plasma due to the large differences in melting and boiling point of constituent elements, associated to the longer resident time of Zn in the plasma. XRF calibration curves for tin and copper are well linear fitted. Consequently also the graph corresponding to copper peak intensity normalized to tin peak intensity, as a function of the element concentration ratio in the samples shows a linear fit with a residual value of the same order for each of the calibration curves. Therefore, in order to compare LIBS results with the calibration curves obtained by XRF analyses, Fig. 10 shows the LIBS analytical copper peak intensity normalized to tin peak intensity, as a function of XRF intensity ratio of the same elements in the samples.

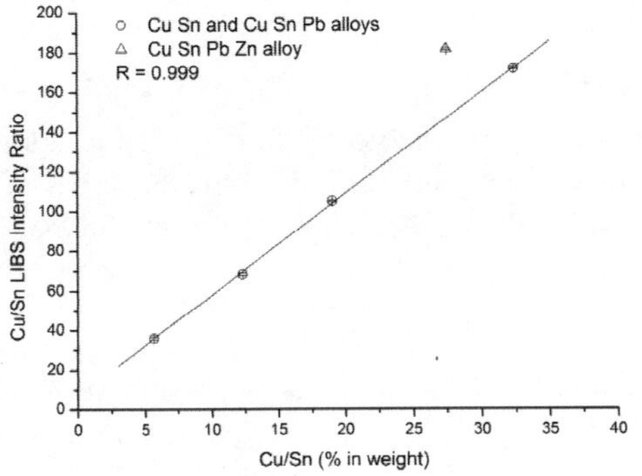

Figure 9. LIBS analytical copper peak intensity normalized to tin peak intensity, as a function of the element concentration ratio in the samples.

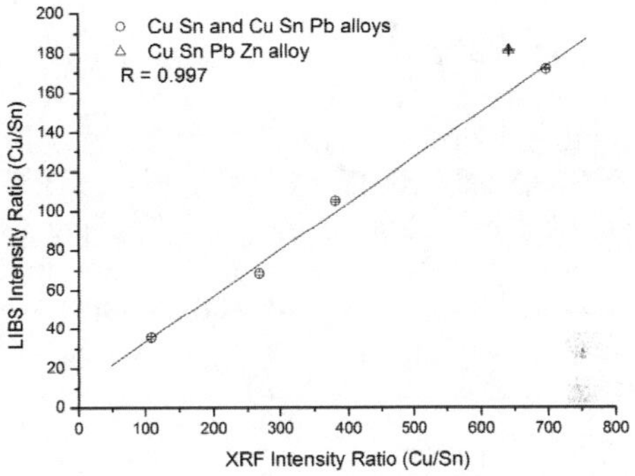

Figure 10. LIBS analytical copper peak intensity normalized to tin peak intensity, as a function of XRF intensity ratio of the same elements in the samples.

The results were in a very well good agreement. Moreover, μ-XRF analysis were carried out on the shot laser point before and after the laser ablation to verify the sample composition. Variations in the composition were not found.

4. Conclusions

The results are highly sensible to the settings of the LIBS experiment, the element to be analyzed, its concentration and the compositional matrix. For what concerns the stability of the shape, morphology and fusion features of the crater on material, the DFT 1 sample has showed a more stable behaviour referable to own thermal-physical features such as heat conductivity, melting and boiling values. The investigations carried out on copper and zinc have allowed to identify the optimal values of the laser pulse energy, the GPW and the GPD time in order to maintain a condition of greater stability in the plasma. Unfortunately, the high fluctuations found in the results related to tin and lead did not permit an optimal assessment of these parameters. It seems appropriate for this purpose, to increase the energy range and the number of spectra accumulated in order to improve the SNR. All these facts lead to the conclusion that, both the adjust of the energy range and the increase of the number of acquisitions could be possible solutions. Nevertheless both involve different problems. The first one causes an increasing of the noise and the second one not is applicable in the cases where it is mandatory to use a non-invasive analytical technique. Then, it is necessary to reach a compromise of both aimed to each particular case study. In order to simultaneously reveal tin and lead and to avoid the above shortcomings, it necessary to reduce the GPD because of the shorter decay life time and to the lower partial vapor pressure of these elements in the plasma. LIBS analytical data relatively to

copper peak intensity normalized to tin peak intensity, as a function of the element concentration ratio in the samples, shown a perfect agreement. In particular they are in good accordance with the analytical data obtained by a well established analytical technique as μ-XRF.
The results presented in this paper, represent our endeavour in order to optimize the best LIBS parameters to be used case by case, in particular taking in consideration applications where the use of micro invasive techniques it's mandatory. Further studies shall be directed to extend our research to other metal alloys, considering also the study of the processes that lead to corrosion phenomena in metals. Moreover, with the scope to better validate the results presented in this paper, our intention will be to apply the Calibration-Free procedure applicable when the LTE conditions are satisfied.

5. Acknowledgments

The authors wish to thank Mr. Marcello Mirabello for their appreciable technical support and Dr. Paolo Guerra for SEM image acquisitions. This research has been carried out within the PRIN 2007 project "Optimization of integrated physical techniques of X-ray spectrometry, laser and electron spin for the study of corrosion processes on surface and volume of metals and alloys", funded by Italian Ministry for University and Scientific Research.

References

[1] Cremers A and Radziemski L J 2006 *Handbook of Laser-Induced Breakdown Spectroscopy* (John Wiley & Sons, Ltd, West Sussex, England)
[2] Miziolek A W, Palleschi V and Schechter I 2006 *Laser Induced Breakdown Spectroscopy (LIBS): Fundamentals and Applications* (Cambridge University Press: Cambridge, UK)
[3] Demtroder W 2008 *Laser Spectroscopy Volume 1 Basic principles* (Springer-Verlag Brlin Heidelberg, Germany)
[4] Tognoni E, palleschi V, Corsi M and Cristoforetti G 2002 *Spectrochimica Acta part B* **57** 1115
[5] Cristoforetti G, Lorenzetti G, Benedetti P A, Tognoni E, Legnaioli S and Palleschi V 2009 *J. Phys. D: Appl. Phys.* **42** 1
[6] Borisov O V, Mao X and Russo R 2000 *Spectrochimica Acta part B* **55** 1693
[7] Fornarini L, Colao F, Fantoni R, Lazic V and Spizzichino V 2005 *Spectrochimica Acta Part B* **60** 1186
[8] Shaikh N M 2006 Plasma diagnostics of Zn, Cd and Hg using Laser-Induced Breakdown Spectroscopy, PhD thesis in Physics, Department of Physics, Quaid-I-Azam University, Islamabad, Pakistan
[9] Borisov O V, Mao X L, Fernandez A, Caetano M and Russo R E 1999 *Spectrochimica Acta Part B* **54** 1351
[10] Gagean M and Mermet J M 1998 *Spectrochimica Acta Part B* **53** 581
[11] Yoonyeol Y, Taesam K, Myeongkwon Y, Kilyong L and Gaeho L 2001 *Microchemical Journal* **68** 251
[12] Pérez-Seredilla J A, Jurado-López A and Luque de Castro M D 2007 *Talanta* **71** 97
[13] Brai M, Gennaro G, Schillaci T and Tranchina L 2009 *Spectrochimica Acta Part B* **64** 1119
[14] Tognoni E, Cristoforetti G, Legnaioli S and Palleschi V 2010 *Spectrochimica Acta Part B* **65** 1
[15] Collins T J 2007 Image J for microscopy *BioTechniques* **43** 25
[16] Griem H R 1964 *Plasma Spectroscopy* Mc Graw Hill Inc.
[17] Reader J and Corlis C H 1980 Part. I Wavelenghts – Wiesse W L and Martin G A Part. II Transition Probabilities in *Wavelenghts and Transition Probabilities for atoms and atomic ions*, National Standard Reference Data System-National Bureau of Standards (Washington, USA)
[18] Handbook of basic Atomic Spectroscopic data, NIST http://www.physics.nist.gov

Journal of Physics: Conference Series **275** (2011) 012018 doi:10.1088/1742-6596/275/1/012018

Influence of successive plasma treatments on PP foils

T. Jacobs*, R. Morent, N. De Geyter, C. Leys

Research Unit Plasma Technology (RUPT), Department of Applied Physics, Faculty of Engineering, Ghent University, Jozef Plateaustraat 22, 9000 Gent, Belgium

* Corresponding author: Tinneke.Jacobs@UGent.be

Abstract. Polypropylene (PP) foil is treated with a dielectric barrier discharge (DBD) plasma operating in helium at medium pressure. The influence of exposure to the atmosphere between successive treatments is studied by varying the exposure time. Each PP sample is treated with subsequent treatment steps of 5 s. Between two treatment steps, different procedures are applied: 1) the sample remains in the discharge chamber at medium pressure (under helium atmosphere) for a certain time before it is treated again or 2) the pressure is increased to atmospheric pressure, so the sample remains exposed to atmospheric air for a certain time and afterwards the system is pumped down again to medium pressure before it undergoes a successive helium plasma treatment. The treated samples are analysed using contact angle measurements. The results show that exposure to the atmosphere between two treatment steps leads to a lower contact angle. The longer the exposure time, the lower the contact angle becomes.

Another experiment showed that the treatment effect could be gradually removed by applying several short plasma treatments of 1 s to saturated samples. With every short treatment step, the contact angle becomes higher. It is believed that this is due to etching of the surface. In the near future, both atomic force microscopy (AFM) and X-ray photoelectron spectroscopy (XPS) analysis on some selected samples are planned to elucidate the chemical and/or physical nature of the observed phenomena.

1. Introduction

Polymers are frequently used in industry for packaging, protective coatings and sealing applications, because they have good bulk properties such as transparency, good thermal resistance, a high strength-to-weight ratio, ... [1]. However, they are often unsuitable to use due to their low surface energy leading to poor wettability, adhesion and printability [2]. Surface modification is often needed. In recent years, plasma treatment of polymers to modify their surface properties has been extensively studied as an environmentally friendly alternative to the classical wet-chemical techniques [3,4]. Different types of plasmas have been used for this purpose [5-9] and several treatment effects have been investigated [10-12]. Since industry requires a plasma technology which can be implemented in a continuous production line, vacuum and medium pressure technology is regarded as being non-competitive with atmospheric technologies. However, medium pressure plasma surface treatment has some advantages over atmospheric pressure. It is easier to create a large plasma volume at medium pressure than it is at atmospheric pressure which could lead to a higher productivity. Also, the pumping equipment for medium pressure is relatively inexpensive. [9-11, 13, 14]

In this study, a dielectric barrier discharge (DBD) operating at medium pressure is used to treat polypropylene (PP) foil. A DBD is typically generated between two metal electrodes, of which at least

Journal of Physics: Conference Series **275** (2011) 012018 doi:10.1088/1742-6596/275/1/012018

one is covered by a dielectric material. By applying a sufficiently large alternating voltage to the electrodes, a plasma is generated in the gap between the two electrodes. Review papers on the properties and applications of DBDs can be found in [15-18].

In this paper, a DBD operating at medium pressure in helium will be used to modify the surface of a PP film. The samples are treated for successive short treatment steps of 5 s and between each plasma treatment step the samples are either kept in the plasma chamber under the helium atmosphere or exposed to atmospheric air. After each treatment step, the plasma treated samples will be examined using contact angle measurements. By using this treatment procedure, the influence of exposure to air between successive helium plasma steps can be studied, which has not been done before.

In a next set of experiments, the saturated plasma treated samples were exposed to several shorter helium plasma treatment steps of 1 s. It was found that the treatment effect could be gradually removed by these shorter treatment steps. This phenomenon has also not been studied before.

2. Experimental set-up

2.1. DBD set-up

A schematic diagram of the DBD configuration can be seen in figure 1. Two circular copper electrodes (diameter = 7 cm) are placed within a cylindrical enclosure. Both electrodes are covered with a glass plate (thickness = 2 mm) and the distance between the plates is 7 mm. The upper electrode is connected to an AC power source (frequency = 5 kHz), the lower electrode to the earth, through a resistor R (50 Ω). The AC power source is a low frequency generator, providing the reference waveform, which is then amplified by a linear amplifier (Crest Audio Model 8200, 4500 W) whose output is applied to the primary winding of a transformer (Tauscher, 2000 VA, 60V/15 kV) in series with a 4.7 Ω resistor. The upper electrode is connected to the secondary of the transformer thereby applying an AC high voltage to the upper electrode.

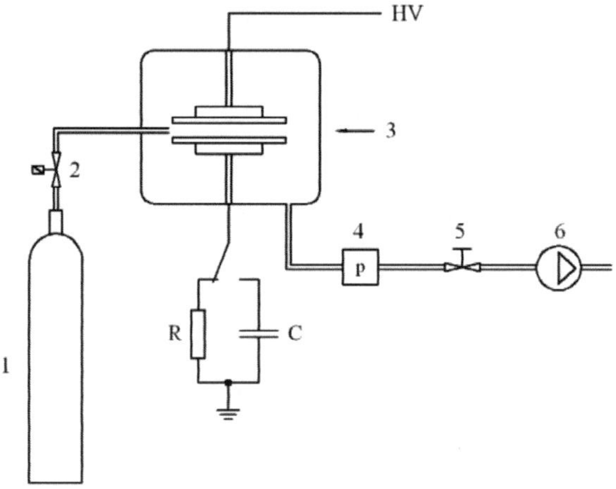

Figure 1: Schematic diagram of the DBD-setup (1. gas cylinder, 2. mass flow controller, 3. plasma chamber, 4. pressure gauge, 5. needle valve, 6. pump).

The discharge current is obtained by measuring the voltage over the resistor R, whereas the applied voltage is measured using a high voltage probe (Tektronix P6015A). The current and voltage waveforms are recorded using an oscilloscope (Tektronix TDS210-60MHz), connected to a computer. The resistor can be replaced by a capacitor (10 nF). The voltage across this capacitor is proportional to the charge stored on the electrodes. Together with the applied voltage, this results in a Lissajous figure. By measuring the area of the Lissajous figure, the electrical power can be calculated [17,18].

Before starting each plasma modification procedure, a PP film with a thickness of 75 μm is placed on the lower glass plate. These commercially available PP foils were purchased from Goodfellow Cambridge Limited (England) and are not subjected to any pre-treatment step before plasma modification. Afterwards, the chamber is pumped down below 2.0 kPa using a rotary vane pump and then filled with helium at a rate of 5.2 slm (standard litres per minute). When atmospheric pressure (101 kPa) is reached, the plasma chamber is pumped down to 5.0 kPa while the gas flow is decreased to 200 sccm (standard cubic centimetres per minute). Subsequently, the AC power source is turned on while the gas flow maintains at 200 sccm and the pressure in the discharge chamber is kept constant at 5.0 kPa by slightly pumping during plasma treatment. In this work, most plasma treatments will be performed in helium purchased from Air Liquide (Belgium) with a 99.999% purity. After this first plasma treatment step, the PP samples will be subjected to different procedures, which will be further explained in detail. Also some treatments in dry air (Alphagas 1 – Air Liquide) will be performed.

2.2. Contact angle measurements

The wettability of the untreated and plasma-treated PP films is evaluated using water contact angle measurements. The contact angles of the foils are obtained at room temperature by using a commercial Krüss Easy Drop optical system (Krüss GmbH, Germany). This system is equipped with a high precision liquid dispenser to precisely control the drop size of the used liquid. A drop of distilled water of 2 µl is placed on the treated substrate. The drop image is then stored, via a monochrome interline CCD video camera, using PC-based acquisition and data processing. Using the computer software provided with the instrument, measurements of the static contact angle are fully automated. The values of the static contact angle are obtained using Laplace-Young curve fitting based on the imaged sessile water drop profile. For each sample, an average of at least 5 measurements was made and standard deviations vary from 0.3° to 1.5°.

3. Results and discussion

3.1. DBD characterization

The voltage and current waveform of the DBD helium discharge is given in figure 2. The amplitude of the applied voltage is approximately 3 kV. Superimposed on the capacitive current, there are several short peaks. These peaks are an indication of the microdischarge activity in the DBD plasma [19]. Every current pulse corresponds to a series of microdischarges.

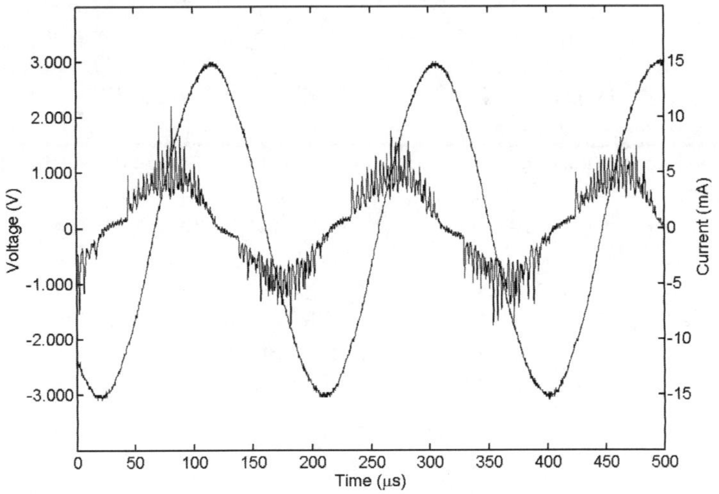

Figure 2: Voltage and current curves of the DBD discharge.

The Lissajous figure, which gives the charge on the electrodes as a function of the voltage over the electrodes, is shown in figure 3. From this figure, an electrical power of 1.08 W can be calculated. This power is kept constant during all measurements.

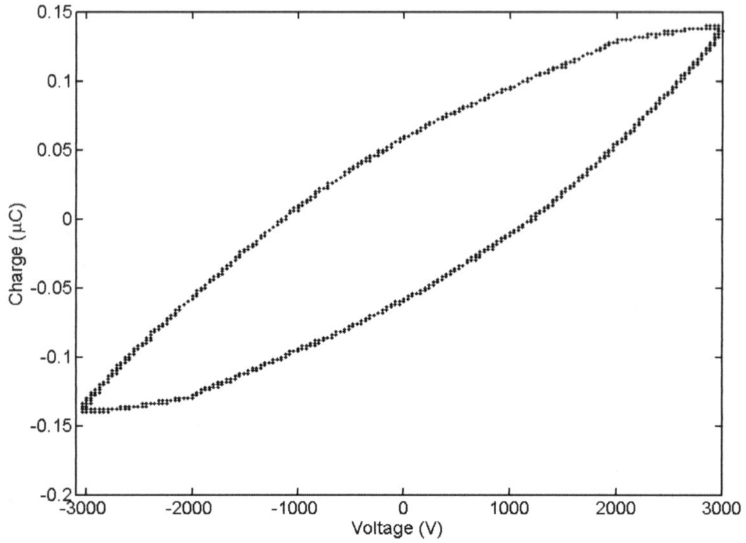

Figure 3: Lissajous figure of the DBD discharge.

3.2. Effect of exposure to the atmosphere in between treatments

To study the influence of exposure to the atmosphere between successive treatment steps of PP foil, three different treatment procedures are performed. A schematic diagram for the three procedures is given in figure 4 and the values for the different parameters are given in table 1. For each procedure, one treatment step consists of exposing the sample to the DBD discharge for 5 seconds ($t_1 = 5$ s). For procedure A, the samples remain in the discharge chamber in helium at medium pressure between two treatment steps ($t_2 = 0.1$ s). For procedure B, the samples were exposed for 10 seconds to the atmosphere by letting air into the discharge chamber ($t_2 = 10$ s). Afterwards, the chamber is pumped down and filled again with helium. For procedure C, the samples were exposed for 5 minutes to the atmosphere between two subsequent treatment steps ($t_2 = 5$ min). The contact angle is measured for each procedure after one, two, three, four and five treatment steps. The contact angle after one treatment step is of course the same for the three procedures. However, after more treatment steps, the contact angle decreases more rapidly for the samples which are being exposed to the atmosphere. This is shown in figure 5. The contact angle also decreases with longer exposure to the atmosphere between two plasma treatments.

It is clear that exposure to the atmosphere has an influence on the treatment, as indicated by the changes in contact angle. By exposing the samples between two treatment steps to atmospheric air, oxygen can be built in. This increase in oxygen content is believed to be due to the reaction of the surface radicals created by the plasma treatment with the oxygen species present in the air. This change in oxygen content at the surface will have an influence on the contact angle. The longer the exposure, the more oxygen can react with the surface radicals and the greater the change in contact angle.

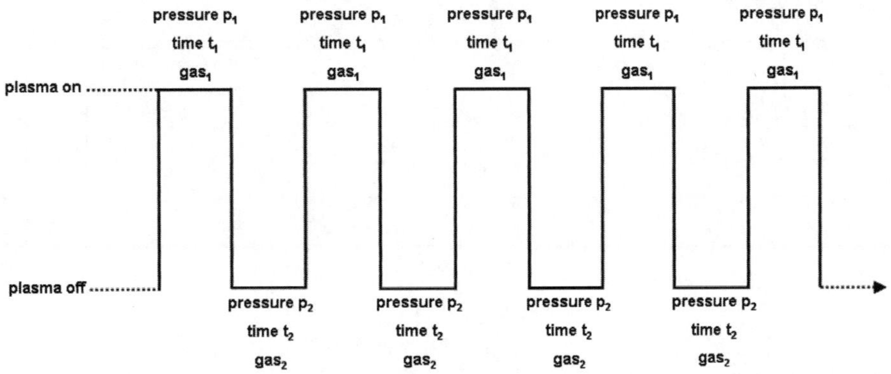

Figure 4: Schematic diagram of the different treatment procedures.

Table 1: Parameters for the different treatment procedures

Parameter	Procedure A	Procedure B	Procedure C
	Treatments in helium		
Pressure p_1	5 kPa	5 kPa	5 kPa
Time t_1	5 s	5 s	5 s
Gas 1	helium	helium	helium
Pressure p_2	5 kPa	100 kPa	100 kPa
Time t_2	0.1 s	10 s	5 min
Gas 2	helium	air	air
	Removal of plasma treatment effect		
Pressure p_1	-	-	5 kPa
Time t_1	-	-	1 s
Gas 1	-	-	helium
Pressure p_2	-	-	100 kPa
Time t_2	-	-	5 min
Gas 2	-	-	air
	Treatments in dry air		
Pressure p_1	5 kPa	-	5 kPa
Time t_1	1 s	-	1 s
Gas 1	dry air	-	dry air
Pressure p_2	5 kPa	-	100 kPa
Time t_2	0.1 s	-	5 min
Gas 2	dry air	-	air

To validate the phenomenon of surface radicals reacting with atmospheric oxygen, plasma treatments in dry air instead of helium were also performed. From the characterization of the discharge it was found that the discharge in dry air was filamentary and the power was 1.83 W. The same procedures A and C as before were used (see figure 4). In table 1, the values for the different parameters are given. Results of contact angle measurements are presented in figure 6. One can see that there is no significant influence of the exposure to the atmosphere in between two treatment steps. During treatment in a dry air plasma, all the created surface radicals immediately react with the oxygen species present in the plasma. No surface radicals are available to react with the oxygen species present in air during exposure to the atmosphere, which means that the surface composition and contact angle do not change.

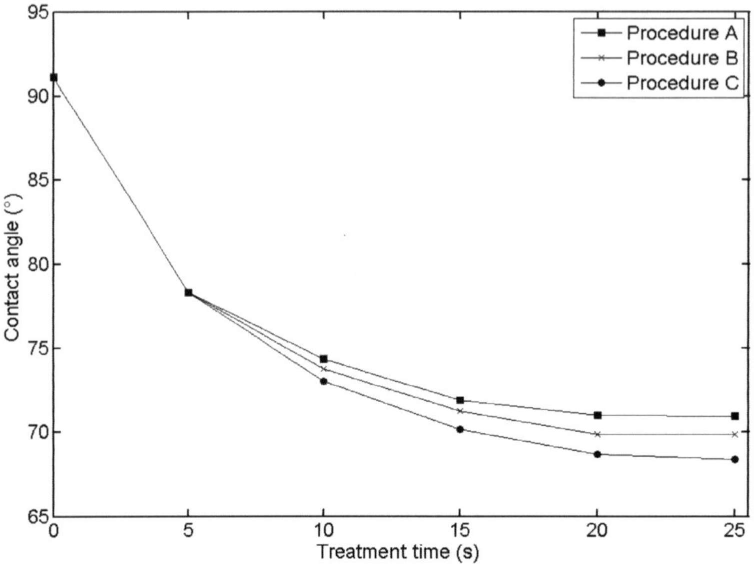

Figure 5: Variation of the contact angle as a function of the treatment time for the three procedures. Procedure A: no exposure to the atmosphere between two treatment steps; procedure B: several seconds of exposure; procedure C: several minutes of exposure.

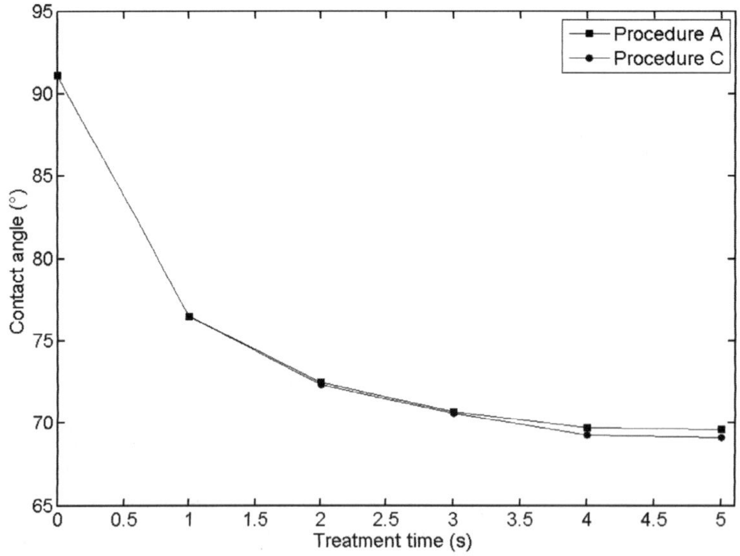

Figure 6: Variation of the contact angle as a function of the treatment time for treatments in dry air.

3.3. Removal of treatment effect

The effect of applying several shorter treatments to a saturated helium plasma-treated sample is also investigated in detail. For the plasma treatments, procedure C is used. The treatment conditions (pressure, gas flow, power, ...) for the short treatment steps are the same as those for the previous longer treatment steps. Only the treatment time (t_1) is changed: 1s instead of 5s (see figure 1 and table 1). Results of contact angle measurements are shown in figure 7. The contact angle gradually increases again, so some of the treatment effect is lost. This change in surface wettability is believed to be due to a decrease in oxygen content. By applying short treatment steps, the oxygen can be etched away from the surface, resulting in a more hydrophobic surface.

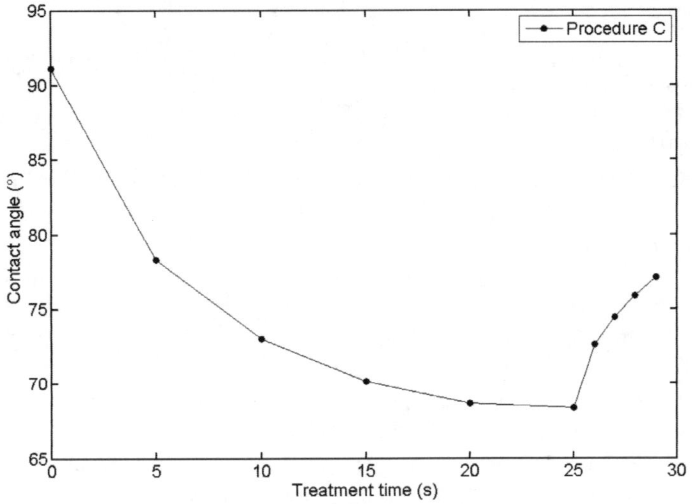

Figure 7: Removal of plasma treatment effect. Procedure C was used to perform the plasma treatments.

4. Conclusion

The results presented in the previous section show that the used DBD set-up is capable of reducing the hydrophobic character of the PP foil and decreasing the contact angle from 91.1° down to 68.4°, in less than 30 seconds of treatment time, at a power of 1.08 W. The exposure to the atmosphere between successive plasma treatment steps in helium at medium pressure has a significant influence on the surface modification. After 5 subsequent plasma steps, the contact angle can vary from 70.9° (no exposure between two successive treatment steps) to 68.4° (5 minutes of exposure). When the same experiments are performed for plasma treatment in air, no difference in contact angle was observed.

When using helium as discharge gas, surface radicals are created at the surface, which can react with oxygen present in air during atmospheric exposure. This increase in oxygen content at the surface will lead to a lower contact angle. When using air as plasma treatment gas, all the surface radicals will react with the oxygen species present in the discharge, leaving no radicals to react with atmospheric oxygen.

When shorter treatment times are performed to a saturated helium plasma-treated sample, the treatment effect can be etched away, and the contact angle increases from 68.4° to 77.1°.

It is believed that both the chemical composition and the surface roughness change during the plasma treatments in helium and air. X-ray photoelectron spectroscopy (XPS) measurements and atomic force microscopy (AFM) measurements on some selected samples are planned to determine the chemical and physical modifications that occur during plasma treatment. This information will make it possible to clarify the observed phenomena.

References

[1] N. Inagaki, K. Narushim, N. Tuchida, K. Miyazaki, J. Polym. Sci., Part B: Polym. Phys. 42 (2004) 3727-3740
[2] O. Kwon, S. Tang, S. Myung, N. Lu, H.S. Choi, Surf. Coat. Technol. 192 (2005) 1-10
[3] A. Kruse, G. Kruger, A. Baalmann, O.D. Hennemann, J. Adhes. Sci. Technol. 9(12) (1995) 1611-1621
[4] S.M. Mukhopadhyay, P. Joshi, J.G. Zhao, P. France, J. Phys. D: Appl. Phys. 35 (2002) 1927-1933
[5] N. De Geyter, R. Morent C., Leys, L. Gengembre, E. Payer, S.Van Vlierberghe, E. Schacht, Surf. Coat. Technol. 202(13) (2008) 3000-3010
[6] H. S. Choi, V. V. Rybkin, V.A. Titov, T.G. Shikova, T. A. Ageeva, Surf. Coat. Technol. 200 (2006) 4479-4488
[7] M. J. Shenton, G. C. Stevens, N. P. Wright, X. Duan, J. Polym. Sci., Part A: Polym. Chem. 40 (2002) 95-109
[8] M. J. Shenton, G. C. Stevens, J. Phys. D: Appl. Phys. 34 (2001) 2761-2768
[9] R. Morent, N. De Geyter, C. Leys, L. Gengembre, E. Payen, Surf. Coat. Technol. 201(18) (2007) 7847-7854
[10] N. De Geyter, R. Morent, C. Leys, L. Gengembre, E. Payen, Surf. Coat. Technol. 201(16-17) (2007) 7066-7075
[11] G. Borcia, N.M.D Brown, J. Phys. D : Appl. Phys. 40(7) (2007) 1927-1936
[12] J.M. Grace, L.J. Gerenser, J. Dispersion Sci. Technol. 24(3-4) (2003) 305-341
[13] R. Morent, N. De Geyter, L. Gengembre, C. Leys, E. Payen, S. Van Vlierberghe, E. Schacht, Eur. Phys. J. - Appl. Phys. 43(3) (2008) 289-294
[14] R. Morent, N. De Geyter, C. Leys, L. Gengembre, E. Payen, Tex. Res J. 77(7) (2007) 471-488
[15] U.Kogelschatz, B.Eliasson, and W.Egli, Journal de Physique IV 7 (1997) 47-66.
[16] U.Kogelschatz, IEEE Transactions on Plasma Science 30 (2002) 1400-1408.
[17] U.Kogelschatz, Plasma Chemistry and Plasma Processing 23 (2003) 1-46.
[18] H.E.Wagner, R.Brandenburg, K.V.Kozlov, A.Sonnenfeld, P.Michel, and J.F.Behnke, Vacuum 71 (2003) 417-436.
[19] U. Kogelschatz, Plasma Chem. and Plasma Process., 23(1) (2003) 1-46

Journal of Physics: Conference Series **275** (2011) 012019 doi:10.1088/1742-6596/275/1/012019

Titanium oxide thin film deposition by pulsed arc vacuum plasma

I S Zhirkov*, C Paternoster, M P Delplancke-Ogletree

Chemicals and Materials, Université Libre de Bruxelles,
50 avenue F. D. Roosevelt, Brussels, 1050, Belgium.

* igzhirkov@gmail.com

Abstract. The experimental study of effects of deposition conditions and plasma parameters on the structure of titanium oxide films is presented. The films are deposited by reactive deposition from a plasma generated by a pulsed cathodic vacuum arc source. The effects of deposition time and substrate bias voltage are studied, the films being deposited at different substrate temperatures. Bragg-Brentano X-ray diffraction and electron microscopy were used to investigate the structure of deposited films. Correlation between plasma parameters, deposition conditions and structure of synthesized films are established.

1. Introduction

Titanium oxides show a great variety of forms and compositions, such as rutile, anatase or brookite [1]. All these forms present the same empirical formula (TiO_2), but their physical and chemical properties are different, due to their crystalline structure. Non-stoichiometric compounds can also be formed in case of an incomplete oxidation of titanium: some of these structures are known as Magnéli phases [2]. For these titanium oxides, the relative amount of titanium and oxygen is expressed by the empirical formula Ti_xO_{2x-1}.

Magnéli phases find applications in tribology and mechanics as economical and effective coatings, since they are solid lubricants [3]. Titanium dioxides show relevant photocatalytic activities, both in the crystalline form and in the amorphous one. This property is useful for purification of water [4], disinfection [5] or for self cleaning coatings [6]. Nevertheless, titanium dioxide, especially in rutile form, is also widely used as oxygen gas sensors [7], in pigments [8], in cosmetic products [9] or optical interference coatings [10].

The growing interest for titanium dioxide and its application as thin films is accompanied by the availability of efficient, versatile and relatively simple deposition techniques, such as pulsed filtered cathodic vacuum arc (FCA, [11]). This method provides many advantages, such as a very high deposition rate [12], a good control of the deposition parameters, and an excellent quality of the coating (very low concentration of defects). In addition, the adhesion [13] of FCA TiO_2 films is good, due to the nearly fully ionized plasma [14]. The structure of titanium dioxide films depends strongly on the deposition conditions: for example, when deposited at room temperature, it can be amorphous [15-17], or crystalline [18-20]. The aim of this work is to understand the relationship between deposition conditions and titanium dioxide film properties; this investigation is in fact needed to achieve a complete control of the film features through a correct tuning of the process parameters.

2. Experimental details

The effect of the substrate temperature (T_{dep}), substrate bias (U_{bias}), deposition time (t_{dep}), arc current (I_{arc}) and working pressure (p_{dep}) were taken into account in this work. The study of the effect of each experimental parameter on the deposited films was considered separately, except for arc current and deposition pressure, whose simultaneous contribution to coating formation was investigated in the same set of samples. The vacuum chamber setup is schematically represented in figure 1. The vacuum arc plasma source and the power supply are described in details elsewhere [21, 22]. Table 1 summarizes the deposition parameters used for the experiments. The grounded cathode is made of pure titanium (99.5% wt). The ion source is operating in the DC-pulsed mode with a maximum arc current of I_{arc} = 400 A, a pulse repetition rate of 3 pps, and a pulse length of 1.5 ms.

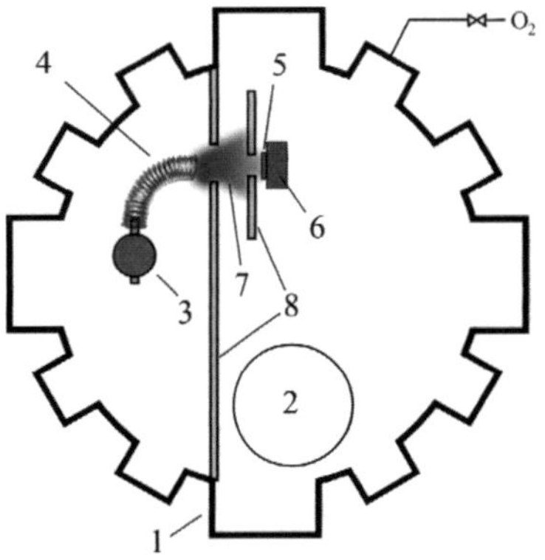

Figure 1. Schematic representation of the deposition chamber: (1) vacuum chamber; (2) cryopump; (3) vacuum arc ion source; (4) magnetic duct; (5) substrate; (6) substrate holder; (7) plasma plume; (8) screens.

The residual pressure in the chamber was 5×10^{-6} mbar. The deposition conditions that were kept constant for all samples are reported in table 1. Pure oxygen gas (O_2 99.5% pure) was introduced at different flow rates in the chamber, to reach, during the experiments, the working pressures reported in table 2. No other gases except oxygen are present, since the deposition is achieved without the use of any sputtering gas (argon, etc.). In the experimental setup used for the deposition, the intake flow of the gas feeding system could be regulated very accurately, to achieve different deposition pressures. On the contrary, the suction line of the vacuum system was not provided with valves, and the pumping rate can only be roughly regulated for pressures less than $\sim10^{-3}$ mbar. Changing the flow rate to achieve different deposition pressures is an approach used by other authors in literature, such as Rosén and coworkers [23].

The plasma plume passed through an electromagnetic filter, for removing the macroparticles ejected from the cathode during the discharge. The distance between the macroparticle filter outlet and the substrate is about 40 mm. The total distance between the cathode and the substrate is about 250 mm. P-doped Si(100) wafers (15x15 mm², 0.6 mm thickness) were used as substrates. A heated substrate holder and a water-cooled one built of copper were used. The heated substrate holder could achieve deposition temperature higher than 773 K, while the other one was used to dissipate as well as possible the heat during deposition at "room temperature" (RT, corresponding at a temperature of about 288 K).

The plasma ionized fraction composition was measured with a RGS 200 (Stanford Research System) gas analyzer, used as a quadrupole-mass spectrometer. The analyzer can identify species with a kinetic energy lower than 12 eV. In the case of a vacuum filtered cathodic arc ion source, the extracted ions can easily reach energy higher than 100 eV, so the energy spectra obtained with this method have to be considered as qualitative. Plasma diagnostic results obtained for different deposition conditions are in accordance with the literature [6, 24 - 26]; Ti^{2+} amount is much higher than Ti^{+} or Ti^{3+} ones. Other species present in small amounts are TiO^{+}, O^{+} and O_2^{+}, and the presence of a small quantity of Fe^{3+}

can be attributed to contaminants. The presence of Ti^{3+} and O^{+} ions could be detected, but the contribution of each species cannot be resolved, because of the same mass-to-charge ratio of these ions.

Table 2 gives the sample names and deposition conditions for each set of experiments.

Table 1. Fixed deposition conditions for all samples investigated in this work.

Base pressure	~ 5 · 10^{-6} mbar	*Pulse duration*	1.5 ms
Working gas	O_2, 99.5% pure	*Pulses per second*	3
	Ti comm. pure,	*Duty cicle*	4.5‰
Target features	Ti = 99.5% wt.	*Cathode-substrate distance*	~ 25 cm
	Ø = 13 mm	*Substrate*	P-doped Si(100)

Table 2. Deposition conditions specifically set for each sample.

Set	Sample name	Substrate temperature *K*	Applied bias *V*	Deposition time *min.*	Arc current *A*	Working pressure $\cdot 10^{-5}$ *mbar*	Flow rate *sccm*
A	A01	773	0	60	400	~ 60	30
	A02	773	0	120	400	~ 60	30
	A03	773	0	240	400	~ 60	30
	A04	773	0	90	400	~ 60	30
B	B01	573	0	60	400	~ 60	30
	B02	573	0	120	400	~ 60	30
	B03	573	0	240	400	~ 60	30
C	C01	423	0	90	400	~ 60	30
	C02	423	-10	90	400	~ 60	30
	C03	423	-50	90	400	~ 60	30
	C04	423	-70	90	400	~ 60	30
D	D01	473	0	240	100	~ 60	30
	D02	473	0	240	100	~ 10	5.3
	D03	473	0	240	100	~ 6	2.3
	D04	288	0	240	100	~ 6	2.3
	D05	473	0	180	200	~ 60	30
	D06	473	0	180	200	~ 10	5.3
	D07	473	0	180	200	~ 6	2.3
	D08	288	0	180	200	~ 6	2.3

To determine the structure of the films, XRD measurements were carried out with a D500 Siemens® diffractometer, equipped with a Cu K_α source ($\lambda_{K\alpha 1}$ = 0.154059 nm). The diffractometer, in the Bragg-Brentano configuration, was operated at 30 mA and 40 kV; data acquisition was set to a stepsize of 0.04° for an acquisition time of 4 sec. for step. The range scan spans from 24° to 64° (2θ angle).
Transmission electron micrographies were obtained with a Philips® CM 20 TEM operating at 200 kV, with a tungsten filament. Scanning electron microscopies were taken with a FEI® Quanta 200 3D, equipped with a tungsten filament; the sample thickness was evaluated from the micrographies.

3. Results

X-rays diffractometry
Set A
The temperature dependence of the TiO_2 films structure has been analyzed, and relevant reflections detected are indexed in the diagram (figure 2 (a)). The reflections at ~33° (2θ) and ~62° (2θ) correspond to the Si(100) substrate [21]. For both T_{dep} = 773 K and T_{dep} = 573 K, the reflections present in the patterns can be attributed entirely to the rutile phase. In particular, for the samples of the examined set, peaks at ~27.4° (110), ~36.1° (101), ~41.2° (111) and ~54.3° (211) are present for the patterns corresponding to all deposition times. The integrated intensity ratios $I_{(101)}/I_{(110)}$, $I_{(111)}/I_{(110)}$ and $I_{(211)}/I_{(110)}$ does not have the same values of the ones of the reference pattern (PDF 021-1276): this is the case of a film growth along preferential directions [21]. The (111) orientation is often the most relevant, independently of the deposition time. New orientations appear after a deposition time of 120 min., at ~44.1° (210) and ~56.6° (220), and they cannot be any more distinguished after a deposition of 240 min., due to the superposition with the peaks corresponding to (111) and (211) planes. No peaks related to other phases have been found for higher deposition temperatures.

Figure 2, (a) and (b), right. XRD diffraction patterns for samples of set A (a) (from bottom to top: A01, A02 and A03) and set B (b) (from bottom to top:B01, B02 and B03)

Figure 3, below. XRD diffraction patterns for samples of set C from bottom to top: samples C01, C02, C03 and C04).

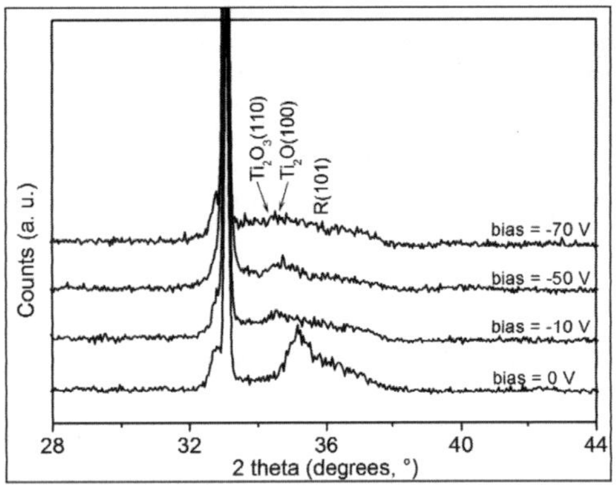

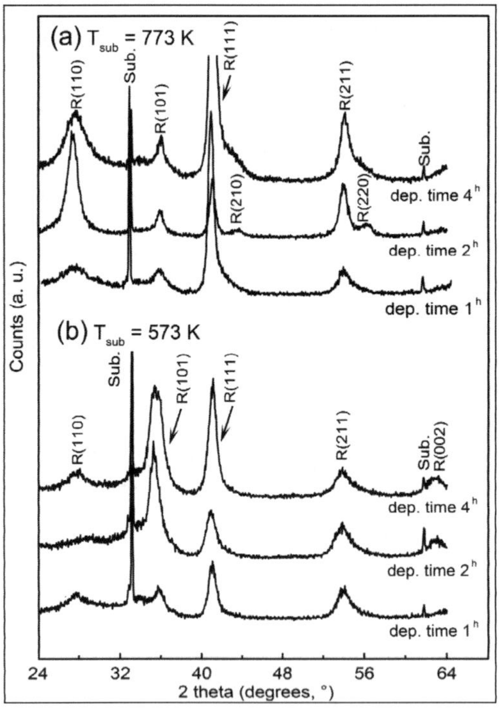

Set B
Several reflections appears in the diffraction patterns of set B samples, but this set shows different preferential growth orientations with respect to the ones already seen for the samples of set A. In this case, in fact, the most relevant peaks are at ~27.4° (101), ~41.2° (111), ~54.3° (211) and ~62.7° (002). The reflection corresponding to the (002) planes appears only for longer deposition times, that is 120 and 240 min.. For all the deposition times, the reflection corresponding to (110), which is the most relevant peak in the PDF reference pattern, is present in the diffractogram only as a broad and weak

peak. On the contrary, the peaks corresponding to (111) and (211) planes are already relevant from the sample with the lowest deposition time; the intensity of these peaks increase however with the deposition time.

Set C

For the deposition carried out at a temperature of 423 K, there is no evidence of a well crystallized phase: the application of a moderate substrate bias, up to -70 V, does not change, in a relevant way, the film structure and features (figure 3). A broad reflection corresponding to the (101) rutile plane is present in all the patterns presented in figure, at a 2θ angle of ~27.4°.

Set D

The diffraction patterns for the TiO_2 coatings are grouped according to the arc current used during the deposition: figure 4 (a) and (b) are representative of the samples deposited with I_{arc} respectively of 100 and 200 A.

The reflections in the diffraction pattern of D01 (p_{dep} = ~ 6 · 10^{-4} mbar, figure 4 (a)) are consistent with the presence of crystallized rutile. The peaks at ~27.4° (101), ~36.1° (101), ~39.2° (200), ~41.2° (111), ~44.1° (211), ~54.3° (211) and ~62.7° (002) do not correspond perfectly to the positions of the reference pattern, and this suggest the presence of residual stresses in the film. A reflection at ~25.3° suggests the presence of a small amount of anatase, with grains oriented in the direction (101). Decreasing the oxygen working pressure induces a decrease of the intensities of many rutile peaks. With a deposition pressure of ~ 6 · 10^{-5} mbar (sample D03), the main rutile peak corresponding to (110) texture is still visible. For the same deposition pressure, and for a deposition temperature of 288 K, the coating is totally amorphous.

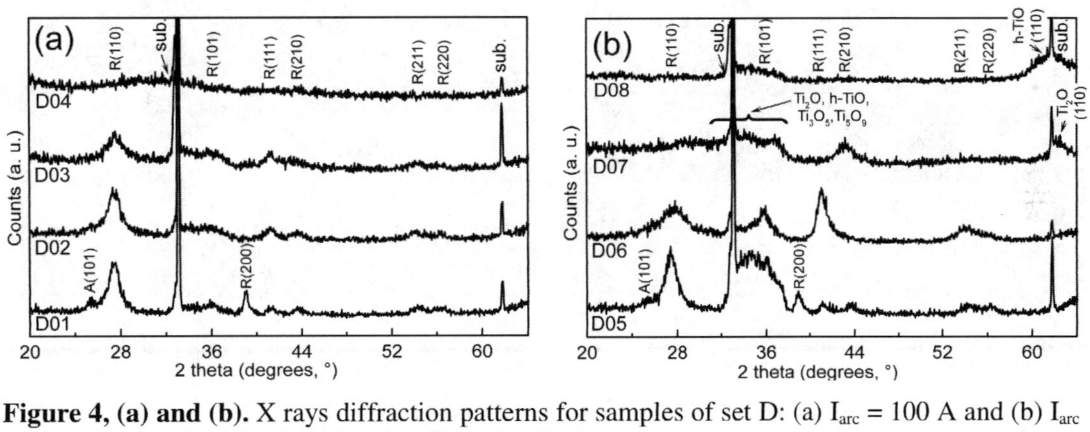

Figure 4, (a) and (b). X rays diffraction patterns for samples of set D: (a) I_{arc} = 100 A and (b) I_{arc} = 200 A. See table 2 for the other experimental details.

For a working pressure of ~ 6 · 10^{-4} mbar, sample D05 deposited with an I_{arc} of 200 A presents reflections corresponding to rutile, the same pointed out for sample D01. In the range ~ 32° to ~ 37°, the irregular shape of the diffraction pattern can be ascribed to the superposition of several peaks, evidence of TiO_x phases with a lower oxygen amount than in rutile. Several small peaks can be attributed to Ti_5O_9 (PDF 011-0193), at ~ 34.4°, ~ 35.7° and ~ 36.2°, related respectively to (-201), (102) and (-1-21) orientations. Other phases, like Ti_2O (PDF 011-0218), h-TiO (PDF 082-0803) or Ti_3O_5 (PDF 040-0806), have reflections in the same angular range. Decreasing the deposition pressure, the intensity of reflections related to rutile also decrease. For deposition carried out at room temperature, only two broad peaks are present, in the range ~ 32° to ~ 37°, and a peak at ~61.1° (h-TiO (110)).

Electron microscopy

TEM micrographs (figure 5) show a sharp interface between the film and the substrate, suggesting coating good uniformity in depth. The structure of the film, for sample A04, present a fine columnar structure. The inset of figure 5 shows the selected area diffraction pattern (SADP). The presence of well crystallized rutile is confirmed for the highest deposition temperatures, and the rings corresponds to several orientations of the crystallites. SEM micrographs (an example is reported in figure 6), here representative of the structure of the film, shows the thickness of the film for sample D07. It was possible to estimate the thickness of the whole set of films: it ranges from few hundreds of nanometers, for most crystallized coatings, to some micrometers for the mainly amorphous ones.
Increasing the deposition time from 60 to 240 min. Induces a thickness *t* increase from ~1.5 to ~6 μm for group A samples and from ~1.3 to ~4.2 μm for group B samples. For group C samples, *t* is ~1.35 for C01, and it ranges between 0.7 ÷ 1.8 for C02 ÷ C04 coatings. In group D, *t* shows the same trend for the two subgroups, even though its value is different for the two subsets: 0.3 to 0.5 μm for subgroup D01 to D04 and 0.5 to 0.9 μm for D05 to D08. In each group, the maximum thickness is reached for a pressure of $\sim 10^{-4}$ mbar.

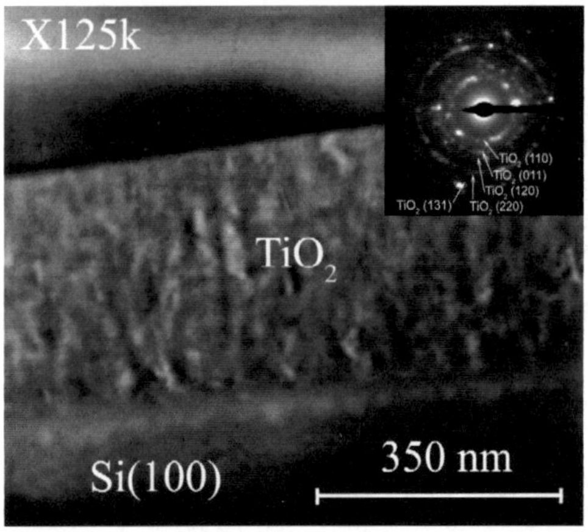

Figure 5. Transmission electron micrograph of sample deposited with the following parameters: T_{dep} = 773 K, $p_{dep} = 6 \cdot 10^{-4}$ mbar, U_{bias} = 0 V, t_{dep} = 90 min.

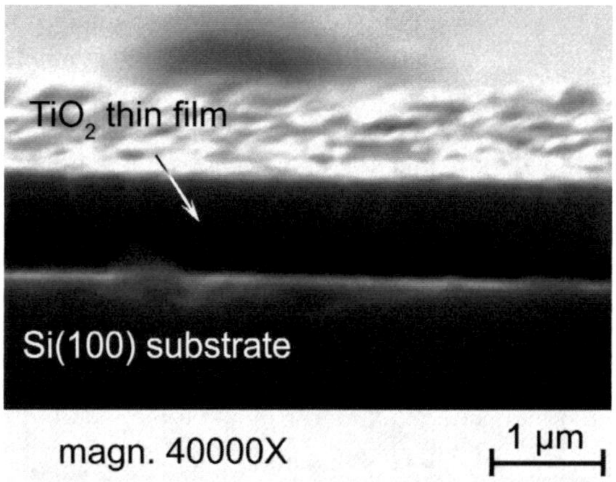

Figure 6. Scanning electron micrograph of sample D07.

Discussion

The rutile phase is formed by the reaction of the Ti^+ and O_2^+ [27], which are present in the plasma. The concentrations of the O_2^+ ions is higher for a lower total pressure, since they are produced through inelastic collisions with high-energy electrons. For lower pressure, the heating responsible for the formation of the rutile phase can be replaced by the energetic of depositing particles. Energetic particles can contribute to the nucleation of rutile crystallites, providing the necessary high mobility of adatoms and insuring the presence of migration effects [20]. Bendavid et al. pointed out also that rutile can also be obtained at room temperature, in presence of an high concentration of ionic species; the presence of such high amount of ionic species is a typical feature of plasma produced by FCA [20].
The ion total energy of the is due to several contributions, highlighted by Anders [28]; ions kinetic, excitation, ionization and cohesive energy, and the energy due to the application of a substrate bias, all play a role, and their relevance may vary with the deposition conditions. The distribution of kinetic energy, as a function of the applied arc current, is reported by Byon and al. [29], but in the range of arc current 100 to 400 A, it does not change in a relevant way. The ionization energies are, on the contrary, significant, especially for multiply charged ions. During FCA deposition, in the case of a bias applied to the substrate, the potential energy can exceed the kinetic energy [28].

Even though electron flux is responsible for chemical modifications of the surface [30], the plasma density is the major factor in the phase transformation [18]; in the present case, the dominant species is Ti^{++}, but Ti^+, TiO^+ O^+ and O_2^+ are also present [22]. The amount of metallic ions ejected by the cathode, and addressed as n_i, i.e. plasma density without the contribution of external gases, is described by the following law: $n_i \propto \alpha I_{arc}$. This relation describes the dependence of the metallic ion density in relation to the applied arc current, with α being the product of several constant, influenced by the deposition parameters and the geometry of the system.

The addition of a working gas changes the typology of ionic species in the plasma, the ion energy distribution [24] and the stoichiometric composition of the film. The overall gas pressure affects also the mean free path of the species of the plasma, so that the energy distribution of the plasma species can vary considerably [24, 31]. For example, in FCA deposition of TiN, Ivanov et al. [24] pointed out that the presence of nitrogen in the deposition chamber decreases r_{Ti} (Ti^{++}/Ti^+ ratio), probably for the enhanced energy transfer from doubly charged energetic Ti^{++} ions with N_2.

When comparing groups A and B samples, we see that the transition between the fully crystallized rutile structure (T_{dep} = 773 K) and the quasi-amorphous phase obtained (T_{dep} = 288 K), is accompanied by a texture change of the deposited film. This texture change can be attributed to phenomena occurring at the surface of the sample, involving adatoms diffusion, nucleation and competitive growth of crystallites. The difference in thickness for A and B groups can be attributed to the different packing of Ti and O; in group B, only (101) and (111) preferred orientations are relevant, while in group A other relevant orientations are present. The prevalence of a specific texture clearly results in closer packing compared with the films without preferred orientation [32, 33].

For sets A and B, no bias was applied, and the only deposition parameter influencing the formation of different phases or states was the substrate temperature: this is responsible for atom diffusion, and rutile crystallite nucleation and growth, and other surface diffusing phenomena. Also in this case the oxygen pressure is sufficient for the formation of TiO_2; the formation of rutile can be attributed to the creation of thermal spikes on the surface, as highlighted by Martin and co-workers [15]: a localized region, at the impact site, is melted and rapidly crystallized after the temperature drop during the following cooling stage. This effect plays an important role when the spike lifetime ($10^{-11} - 10^{-12}$ s) is higher than the lifetime of the collision cascade [18].

The results shown in Figure 3 are representative of the effect of a substrate bias. It is responsible for the formation of oxygen-deficient titanium oxides, TiO_x with $x < 2$. The films deposited at even lower temperatures are amorphous. The thermal energy provided through the heating of the substrate in this case is too low to induce a consistent crystallization of the coating [34]. We detect the presence of a broad rutile peak, and of a peak attributed to an oxygen-deficient titanium oxides phase. This can be due a dominant ion bombardment effect. The increased energy provided from the bias to the ions can give a contribution for TiO_x with $x < 2$. The thickness evolution for group C can be attributed to the concurrent phenomena of resputtering, more evident for C02, and of the increased ion flow for higher applied bias (sample C04).

The thickness variation in group D is due to the increased amount of Ti ions available with higher arc currents; these energetic ions impinging on the substrate also contribute to change the structure of the coating, making it amorphous, and increasing consequently its thickness [12, 32, 33]. For this set of samples, when a I_{arc} = 100 A is applied, the formation of a crystallized rutile is observed, even for a temperature of 473 K and an oxygen pressure of $\sim 6 \cdot 10^{-4}$ mbar, while at the same pressure and temperature, with a I_{arc} = 400 A is applied at the cathode, there is no crystallized rutile [21]. Decreasing the oxygen partial pressure in samples D01 ÷ D04 does not prevent the rutile crystallization, even though the reflection intensity decreases, and reflections due to the presence of poorly crystallized phases are still present. For samples D05 to D08, substoichiometric oxides are present when the pressure is decreased. This behaviour can be attributed to the higher Ti ions amount in D05 to D08 samples than in D01 to D04, because of the higher arc current applied [18]. On the other side, the working pressure is an important parameter in filtered cathodic vacuum arc (FCVA) deposition, because an increasing pressure of the process gas is responsible for decreasing the average charge state and the average energies [23, 25, 35]. When I_{arc} is increased up to 400 A, the coating is almost totally amorphous (sample C01). With the same parameters, that is I_{arc} = 400 A and $p_{O2} = \sim 6 \cdot 10^{-4}$ mbar, Zhirkov and coworkers [22] deposited a set of sample, with a cristallinity degree increasing with increasing temperature. In these conditions, Ti and O amounts are sufficient for the formation of rutile.

According with XRD patterns for samples D01 and D05, crystalline rutile was achieved also with lower arc currents. This means that for I_{arc} = 100 to 200 A, the involved ion energy is somewhat higher, allowing the crystallization of rutile and of a small amount of anatase.

This outcome is apparently in contrast with results presented by other authors. Bendavid and coworkers [18] states that a critical ion current density (CICD) is needed for a phase transformation. The crystalline phase is formed for higher ion current densities than the critical one, which was 0.5 mA/cm^2. CICD depends on plasma ion density, which is related to the arc current used for deposition [18]. New investigations and experiments are needed to better understand the phenomena which contributed to this particular result (Ti ions - reactive O gas interaction, charge state and ion energy plasma distribution, substrate resputtering, etc.).

Conclusions

Thin titanium oxide films were deposited with a filtered cathodic arc system. Effects of substrate temperature and bias were investigated in the present work. Rutile/amorphous transition was achieved, by means of thermal energy and without any substrate biasing. A rutile structure is obtained for temperatures higher than 773 K and an amorphous phase for temperatures below 423 K.

Different mechanisms can be responsible for the structure of the films: (1) thermal energy, provided by substrate heating, (2) energetic ion bombardment from the plasma, (3) arc ion current, responsible for the amount of titanium ejected from the cathode, and (4) amount of oxygen. Thermal energy and substrate bias can affect the properties of the coating; in particular, for a t_{dep} = 473 K and a I_{arc} = 100 A well crystallized rutile was obtained, while for the same temperature, with an I_{arc} = 400 A poorly crystallized films were obtained. By decreasing the pressure, well crystallized rutile was obtained even for low arc currents.

The formation of titanium oxides like Ti_2O or Ti_2O_3, for low temperature and substrate biasing up to -70 V, can be attributed to the synergy of several phenomena, such as resputtering, bombardment of metallic energetic ions and reduced atom mobility due to the low amount of thermal energy. The mechanisms involved in the formation of amorphous films, even for a high ion current density corresponding to I_{arc} = 400 A, has to be clarified. Further experiments needs to be carried out, taking into account different deposition parameters, to better understand the formations of the preferred orientations at higher temperature deposition.

Acknowledgements

The research was performed as part of the “Interuniversity Attraction Poles” IAP-PAI program financed by the Belgian government (BELSPO). The authors would like to thank the EXCELL Network of Excellence (NMP3-CT-2005-515703) for its support.

References

[1] Diebold U, 2003 *Surf. Sci. rep.* **48** 53
[2] Langlade C, Vannes B, Sarnet T and Autric M, 2002 *Appl. Surf. Sci.* **186** 145
[3] Storz O, Gasthuber H and Woydt M, 2001 *Surf. and Coat. Tech.* **140** 76
[4] Mills A, Davies H R and Worsley D, 1993 *Chem. Soc. Rev.* **22** 417
[5] Maness P C, Smolinski S, Jacoby W A, 1999 *Appl. Environ. Microbio.l* **65** 4094
[6] Paz Y, Luo Z, Rabenberg L and Heller A, 1995 *J. Mater. Res.* **10** 2842
[7] Sheveglieri G (ed.), 1992 *Gas sensors* Dordrecht : Kluwer Academic Publisher
[8] Ryu Y C, Kim T G, Seo G-S, Park J H, Suh C S, Park S-S, Hong S-S and Lee G D, 2008 *J. of Ind. and Eng. Chem.* **14** 213

[9] Hewitt G 1999 *Cosmet. Toiletries* **114** 59
[10] Yang C, Fan H, Xi Y, Chen J and Li Z, 2008 *Appl. Surf. Sci.* **254** 2685
[11] Anders A, 2000 Handbook of Plasma Immersion Ion Implantation and Deposition, 1st ed., Wiley-Interscience
[12] Bendavid A, Martin P and Preston E, 2008 *Thin Solid Films* **517** 494
[13] Boxman R L, Haber D and Martin P J, 1997 Handbook of Vacuum Arc Science & Technology: Fundamentals and Applications, William Andrew
[14] Martin P J, Bendavid A, Netterfield R P, Kinder T J, Jahan F and Smith G, **1999** *Surf. and Coat. Tech.* **112** 257
[15] Takikawa H, Matsui T, Sakakibara T, Bendavid A and Martin P J, 1999 *Thin Solid Films* **348** 145
[16] Zhao Z W, Tay B K, Lau S P, Yu G Q, 2004 *J. of Cry. Growth* **268** 543
[17] Zhang M, Lin G, Dong C and Wen L, 2007 *Surf. and Coat. Tech.* **201** 7252
[18] Bendavid A, Martin P and Preston E, 2008 *Thin Solid Films* **517** 494
[19] Bendavid A, Martin P J, Jamting Å and Takikawa H, 1999 *Thin Solid Films* **355 - 356** 6
[20] Bendavid A, Martin P J and Takikawa H, 2000 *Thin Solid Films* **360** 241
[21] Chekh Y, Zhirkov I S and Delplancke-Ogletree M P, 2010 *Rev. Sci. Instrum.* **81** doi:10.1063/1.3310096
[22] Zhirkov I S, Paternoster C, Chekh Y and Delplancke-Ogletree M P, *Proceedings of the First European Conference on NanoFilms (ECNF 2010)*, March 22-25, 2010 - Liège, Belgium
[23] Rosén J, Anders A, Mráz S, Atiser A and Schneider J M, 2006 *J Appl Phys* **99** 123303-1
[24] Ivanov I, Ljungcrantz H, Håkansson G, Petrov I and Sundgren J-E, 1997 *Surf. and Coat. Tech.* **92** 150
[25] Tarrant R N, Bilek M M M, Oates T W H, Pigott J and McKenzie D R, 2002 *Surf and Coat Tech* **156** 110
[26] Anders A, 1997 *Phys Review E* **55** 969
[27] Okimura K, Shibata A, Maeda N, Tachibana K, Noguchi Y and Tsuchida K, 1995 *Jpn. J. Appl. Phys.* **34** 4950
[28] Anders A, 2002 *Vacuum* **67** 673
[29] Byon E and Anders A, 2003 *J. of Appl. Phys.* **93** 1899
[30] Konstantinidis S, Dauchot J P and Hecq M, 2006 *Thin Solid Films* **515** 1182
[31] Okimura K, Naohiro M and Shibata A, 1996 *Thin solid films* **281-282** 427
[32] Aarik J, Aidla A, Kiisler A-A, Uustare T and Sammelselg V, 1997 *Thin Solid Films* **305** 270
[33] Ottermann C R and Bange K, 1996 *Thin Solid Films* **286** 32
[34] Mändl S, Thorwarth G and Rauschenbach B, 2000 *Surf and Coat Tech* **133-134** 283
[35] Bergman C, 1988 *Surf and Coat Tech* **36** 243

Journal of Physics: Conference Series **275** (2011) 012020 doi:10.1088/1742-6596/275/1/012020

Styrene and methyl methacrylate copolymer synthesized by RF inductively coupled plasma

Z. Li, X. Gillon, M. Diallo, L. Houssiau, J. -J. Pireaux
University of Namur (FUNDP)
Research Centre in Physics of Matter and Radiation (PMR),
61, Rue de Bruxelles, 5000 Namur, Belgium

E-mail: zhiling.li@fundp.ac.be

Abstract. A series of random copolymers of styrene and methyl methacrylate was prepared on a silicon substrate by RF pulsed inductively coupled plasma. The plasma gas phase was investigated by optical emission spectroscopy (OES). The physico-chemical characteristics of the deposited copolymer films were analyzed by several surface techniques: X-ray photoelectron spectroscopy (XPS), Fourier-Transform infrared absorption (FT-IR), Time-of-flight secondary ion mass spectrometry (ToF-SIMS), etc. OES of the plasma and FT-IR spectra of the films are predictive: plasma emitting a higher relative benzyl radical signal results in the deposition of a more aromatic plasma polymer. The functional thin films can be deposited by selection of the co-monomers.

1. Introduction

Plasma copolymerization is a relatively new method for the synthesis of polymer materials, which have diverse applications in medicine and engineering. It was proved that by combining the properties of different monomers, plasma copolymerization have the ability to design functional thin films [1]. For decades, many studies have been focused on the polymerization and characterization of films formed by the plasma from a single precursor or a monomer and different noble gases [1-2]. Fewer reports discussed the process of copolymerization by plasma [3-5], still fewer works aimed to diagnose the plasma copolymerization process.

Compared to conventional chemistry, the reactions in the plasma are much more complex. In plasma copolymerization, the initiation reactions are based on the collision probabilities between different species (monomer molecule, electron, ion, excited species, free radicals, etc.). The injection of a second monomer adds to the complexity of the reactions among these species. These reactions can be instantaneous, or proceed stepwise; they depend on the selected co-monomers and of the parameters of the plasma process. [2-6]

We studied the effect of power, pressure, deposition time and duty cycle of radiofrequency (RF) pulsed plasma on the polymerization of styrene and methyl methacrylate (MMA). In this paper, we study that for different ratios of monomers in the plasma copolymerization process, we investigated the content of species in the plasma itself and the composition of the deposited thin films.

Journal of Physics: Conference Series **275** (2011) 012020 doi:10.1088/1742-6596/275/1/012020

Polymethylmethacrylate (PMMA) is widely used as a sheet glazing material, optical lens and fluorescent solar collectors, because of its optical clarity and UV resistance. Polystyrene has reasonably good mechanical and thermal prosperities with its high refractive index, but its stability to outside weathering is poor: it becomes yellow easily on exposure. The optical properties of a cyclic olefin copolymer are attractive on account of their high use temperature, excellent optical transmission, low birefringence and low moisture uptake. The copolymerization by plasma could overcome the defects of homopolymers. [7]

In the literature, most of the optical emission spectroscopy (OES) reports cover a very limited spectral range and only the strongest lines are identified; to our knowledge, very few detailed OES analyses of co-monomer plasma have been performed. We expect, and indeed show, that much more information can be extracted from the detailed analysis of full emission spectra recorded from the UV-visible to the near infrared spectral range.

To complement this information, the copolymer films prepared by RF plasma were characterized by multiple techniques: X-Ray Photoelectron Spectroscopy (XPS), Time of Flight Secondary Ion Mass Spectrometry (ToF-SIMS), and Fourier-Transform Infrared spectroscopy (FT-IR). From a correlation of the information gained from the composition of the plasma phase, and the analyses of the deposited copolymer films, we will infer some mechanisms of the plasma copolymerization.

2. Experimental

The samples were prepared in a stainless steel chamber which was pumped at 0.1Pa. Plasma was generated in an inductively coupled planar radio frequency (RF) discharge (13.56MHz, ICP-P 200, JE PlasmaConsult GmbH, Germany). The reactor design is reported elsewhere [8].

The distance between the quartz window of the plasma source and the sample support is 12cm. Deposition time is 8min in each case. Styrene and MMA monomers were obtained from Acros Organics (Geel, Belgium). Copolymers of styrene and MMA were prepared onto a clean silicon substrate at a fixed plasma power of 120W, duty cycle 50% (T_{on}=0.5ms, T_{off}=0.5ms), for a total reactor pressure of 25Pa.

Each monomer gas is injected by a transport tube terminated by a "shower ring" output, allowing a homogeneous gas distribution in the discharge region of the chamber. A pressure meter (MKS-Baratron) measures the actual pressure in the chamber. After the pressure of the reactor went down to 0.1Pa, the first monomer was injected to reach the desired pressure, then, the second monomer vapor was injected. When the total pressure reached the set value of 25Pa, it took 15min to wait until the pressure was stable. A summary of the experimental conditions used in the plasma copolymerization is given in Table 1.

Table 1 Experimental compositions of the styrene and MMA plasma at 120W plasma power and 25Pa.

$P_{st}/(P_{st}+P_{MMA})$ %	MMA Pa	Styrene Pa
0	0	25
20	5	20
40	10	15
50	12.5	12.5
60	15	10
80	20	5
100	25	0

It is worth noting that if an ideal gas behaviour is assumed, the $P_{st}/(P_{st}+P_{MMA})$ ratio, where P_{st} and P_{MMA} are the monomer pressure of styrene and MMA, respectively, is also the mole fraction of styrene in the gas feed.

OES measurements were performed with a spectrometer equipped with a 3-grating monochromator (Princeton Instruments SP2756A, f/9.7). The light emitted by the plasma was collected through a

Journal of Physics: Conference Series **275** (2011) 012020 doi:10.1088/1742-6596/275/1/012020

quartz window, by an optical fibre, and analyzed onto a cooled CCD camera. The spectral response of the CCD detector has been calibrated using a reference Halogen lamp (LSK114, LOT Oriel). The calibration of lines is executed by the Hg (Argon) lamp (LSP035, LOT Oriel)

The chemical composition of the deposited polymer surfaces was investigated by XPS (Surface Science SSX-100), by FT-IR spectroscopy (FT-IR, BIO-RAD 60A), and by ToF-SIMS (ToSIMS IV, ION-TOF GmbH). The deposition rate is obtained from the measurement of the polymer film thickness using a Dektak profilometer.

3. Optical emission spectroscopy of the plasma phase

Optical emission spectra measured in the 200 and 1000nm wavelength range. The spectra are dominated by all kinds of atomic lines and molecular bands of the plasma gas which are related to the monomers used in the plasma (styrene, MMA, and mixture of both monomers).

The three spectra in Figure 1 correspond to plasma gas of pure styrene, of the 80% MMA-20% styrene mixture, and of the pure MMA, respectively. The OES spectra recorded in pure MMA and pure styrene plasma differs completely. The spectrum of MMA plasma is composed of more lines, and they are more intense compared with that of styrene plasma. Most of the atomic lines and molecular structures have been identified by using literature references [9-12], and the excited species in the plasma show as being very different of that of styrene plasma. All the experimental spectra related to the styrene show a strong blue continuum between 260nm and 350nm with a maximum intensity around 300nm, while the spectrum from the MMA plasma shows a continuum between 250nm and 800nm with a maximum intensity around 400nm.

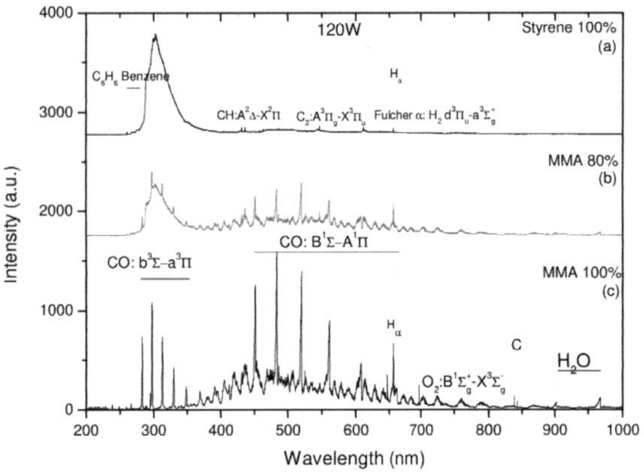

Figure 1 OES from the plasma gas of MMA monomer, MMA 80% and styrene 50% monomers, and styrene monomer at 120W, DC50%.

In the OES of the pure styrene plasma (Figure 1(a)), these excited species are clearly seen: C_6H_6, CH, C_2, C, H, and H_2. Among them, the most intense molecular band observed is the ro-vibration transition (0-0) at 431 nm in the $A^2\Delta$-$X^2\Pi$ of the CH signature, and its higher vibrational level (2, 2) at 432nm. The continuum (260 to 350nm) corresponds to the transitions from the lowest stable triplet state ($a^3\Sigma^+_g$) of the H_2 molecule to the repulsive state ($b^3\Sigma^+_u$). 266.7nm, 273.9nm and 260.2nm are the vibrational band heads of the benzene molecular emission which superimpose onto the H_2 emission continuum. The molecular bands at 509nm, 512nm, and 516.2nm could be associated with the band heads of the 3 different C_2 vibrational levels of $d^3\Pi_g$-$a^3\Pi_u$ electronic transition, known as the "swan system" [10][13]. Moreover, a weak continuum between 420 and 600nm is attributed to the hot incandescent carbon particles produced in the discharge [14]. The Fulcher-α band around 600nm corresponds to transition between the $d^3\Pi_u$-$a^3\Sigma^+_g$ electronic states. The H_α atomic line (656.2nm) formed in the styrene plasma is very intense. The other hydrogen atomic lines (434nm and 486nmnm) can be observed in higher plasma power and at lower pressure. No optical emission bands or atomic lines from oxygen were detected in the wavelength range of the optical emission system, so the representation of the H_2 molecules can directly be explained from the combination of hydrogen atom in the plasma gas. The carbon atomic line at 887.3nm and 889.9nm observed are weak.

Journal of Physics: Conference Series **275** (2011) 012020 doi:10.1088/1742-6596/275/1/012020

In the MMA plasma (Figure 1.c), at least ten systems of carbon monoxide dominate the emission spectrum (Table 1) on the whole range of wavelengths including the Angstrom system $B^1\Sigma_{2u}$-$A^1\Pi_{1g}$, the third positive system $b^3\Sigma_{2u}$-$a^3\Pi_{1g}$ and the Asundi system $a^3\Sigma_{2u}$-$a^3\Pi_{1g}$ [15] which is superimposed onto a continuum emission due to the chemiluminescence accompanying the recombination of CO and O [16]. In the 800-1000nm range, the O_2 and H_2O molecules can also be identified, but with a weaker intensity. The complex secondary spectrum of H_2 Fulcher-α appears too, while the emission of CH at 431.4nm (established using a higher resolution) is very weak: it is quite difficult to observe.

Table 2 Species identified in MMA and Styrene plasmas.

Species in styrene plasma			Species in MMA plasma		
Species	Configuration	Wavelength (nm)	Species	Configuration	Wavelength (nm)
C_6H_6	A^1B_{2u}-X^1A_{1g}	266.7, 273.9, 260.2	CO	$b^3\Sigma_{2u}$-$a^3\Pi_{1g}$ (0→0)	283.3
CH	$A^2\Delta$–$X^2\Pi$ (0→0)	431		(0→1)	297.3
	(2→2)	432		(0→2)	312.7
C_2	$d^3\Pi_g$-$a^3\Pi_u$(0→0)	516.5		(0→3)	329.7
	(1→1)	512.5		(0→4)	348.5
	(2→2)	509.7		(0→5)	369.4
H_2	Fulcher α: : $d^3\Pi_u$-$a^3\Sigma^+_g$	570-650		$B^1\Sigma_{2u}$-$A^1\Pi_{1g}$(0→0)	451.0
Atomic line				(0→1)	483.5
C	5s1P-p3D	887.3		(0→2)	519.8
	8p1D-3d1D	889.9		(0→3)	561
H	H_β 4d-2p	434		(0→4)	607.99
	H_γ 5d-2p	486.1		(0→5)	662.03
	H_α 3d-2p	656.2	Atomic line		
			H	H_β 4d-2p	434
				H_γ 5d-2p	486.1
				H_α 3d-2p	656.2
			O	3p5P-3s5S	777.2

Adding the styrene monomer into the MMA plasma, the OES changes drastically, but this emission spectrum has superimposed both the characteristic peaks of styrene and MMA plasma (Figure 1 (b)), including the CO systems, the molecular bands CH, C_6H_6, H_2, and O_2, the two strong continuums and the atomic lines H, O. The several systems do overlap each other, so high resolution OES is required to confirm the identification of these species. At least ten systems of CO from the composition of MMA are identified; the emission from H, H_2, CH and C_2, in the 350-850nm part of the OES, superimpose onto the different systems of CO. A continuum of 400-700nm observed in the OES of MMA plasma is less intense than that in the pure MMA plasma and the hydrogen continuum of 200-350nm are also visible. Only when the styrene content added into the co-monomer plasma is enough high, it is possible to observe the C_2 band.

When the relative fraction of the two monomers is varied (for a fixed total pressure: p=25Pa), the characteristic peaks corresponding to the different monomer plasma evolve differently, particularly in the CO system. A high resolution spectrum is shown as a function of the MMA content in Figure 2. The emission intensity of excited species (CO, C_6H_6 benzyl radical, etc.) correlates with the concentration of the monomer in the plasma. The CO emission peak is very intense even after the small amount of quantitative MMA was added to the plasma. In the other hand, the H_2 continuum around 300nm became increasingly weak and the H atomic line at 656.2nm compared with the CO peaks is also weak. As for the CO and C_6H_6 band intensities, one notes (Figure 3) that the CO emission peak intensity increased and the C_6H_6 benzyl radical decreased as the MMA monomer increased.

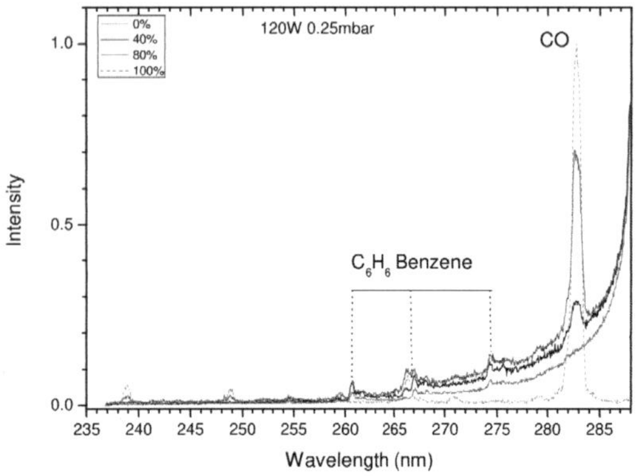

Figure 2 Spectral evolution of the C_6H_6 benzene and CO molecular bands as a function of MMA fraction.

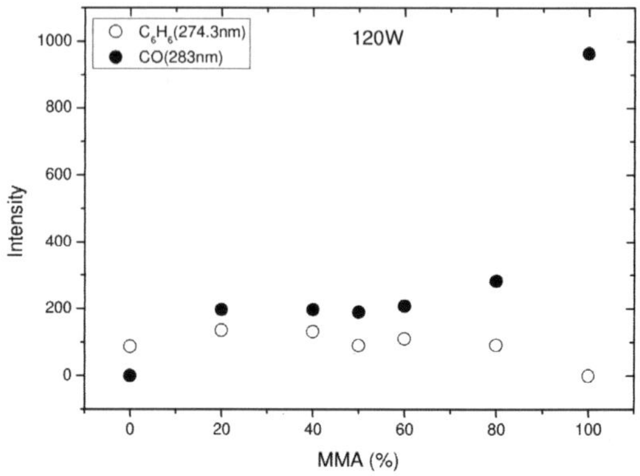

Figure 3 Intensity of the CO and C_6H_6 benzene bands as a function of the MMA fraction.

4. Physico-chemical characteristics of the deposited copolymer films

Figure 4 (a) presents the curve-fitted high resolution XPS spectrum in the C1s region for the plasma deposited MMA (PMMA) surface. Three different carbon environments are found, as expected: carbon bonds at 284.5eV, C-O bonds at 285.5eV, C=O bonds at 288.5eV. With a small fraction of styrene, the plasma copolymer surface immediately shows the characteristics of a plasma-polystyrene surface, with the fingerprinting π–π* transition peak at 291eV (Figure 4 (b)); the intensity of this shake-up transition increases with the styrene fraction in the plasma discharge. The styrene XPS signals dominate the XPS spectrum for each deposited copolymer. Similar behaviour can be observed in the evolution of the copolymer deposition rate (Figure 5). One notes that, although the addition of styrene is very small, the deposition rate dramatically increases, while the XPS spectrum of the deposited material becomes dominated by the styrene component. The styrene monomer has the function of an accelerator for the deposition rate.

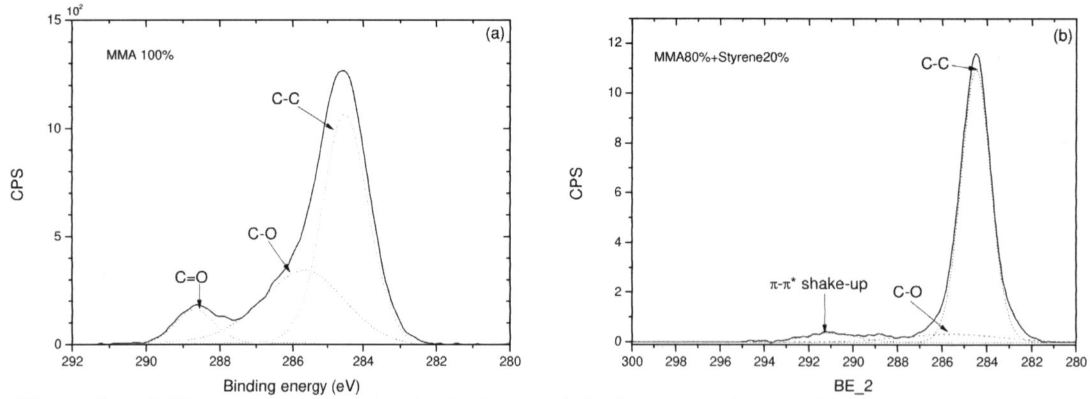

Figure 4 XPS C1s spectra for 'pure' polyMMA (a) and (b) for a MMA/styrene 80-20% copolymer system.

Figure 4 suggests that the surface aromaticity of MMA and styrene copolymer films increases with the addition of styrene in plasma. This behavior is consistent with infrared spectra (Figure 6). The C-H stretch band at 3100-3000cm^{-1} is characteristic of films which contain aromatic moieties. Figure 6 shows that the peak height of the aromatic C-H stretching vibration at 3027cm^{-1} increases by comparison to that of the asymmetric methylene C-H stretching vibration, with addition of styrene. The corresponding aromatic to aliphatic absorption intensity ratios obtained from the IR spectra are shown Figure 7. More aromaticity was observed in the deposited copolymer films, in agreement with the OES of co-monomer plasma. On the contrary, the CO_2 asymmetric and the C=O carbonyl bands decrease.

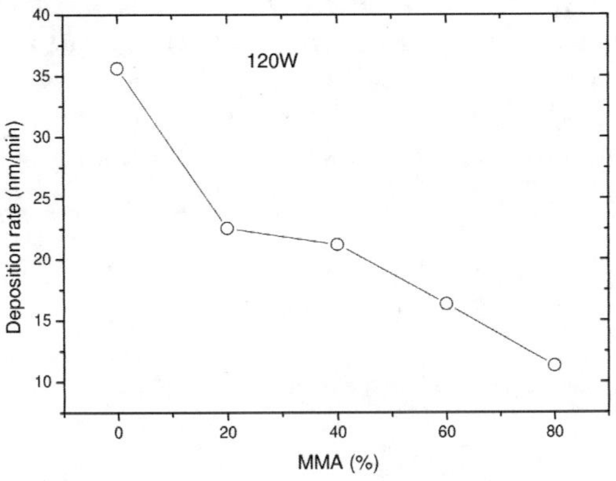

Figure 5 Deposition rate as a function of MMA fraction.

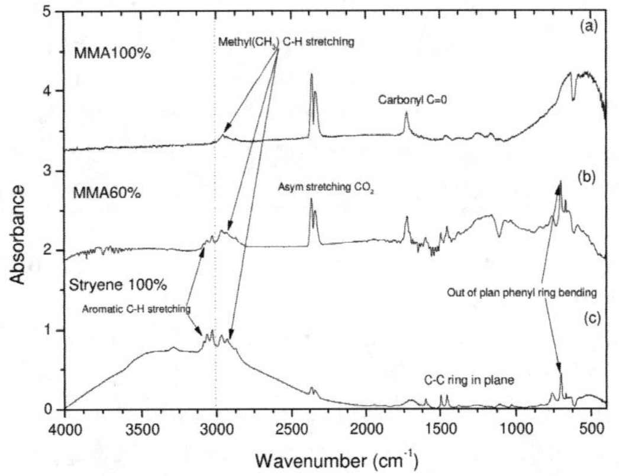

Figure 6 IR spectra of deposited films for different MMA fraction.

Surfaces of several styrene–MMA copolymer films have been analyzed by ToF-SIMS in order to detect any possible surface segregation of one of the two components. From the XPS data, one has indeed noticed that a small fraction of styrene content in the plasma gas resulted in a material with a surface showing essentially a styrene-like content (Figure 4 (b)). SIMS spectra of styrene based polymers are mainly composed of hydrocarbon clusters with unsaturated and/or aromatic fragments ($C_2H_2^+$, $C_2H_3^+$, $C_3H_3^+$, $C_4H_3^+$, $C_5H_3^+$, $C_6H_5^+$, $C_7H_7^+$, $C_8H_7^+$, $C_8H_9^+$, $C_9H_7^+$, $C_{13}H_9^+$, and $C_{15}H_{13}^+$ at *m/z* 26, 27, 39, 51, 63, 77, 91, 103, 105, 115, 165, and 193, respectively). Figure 8(a) shows these polystyrene characteristic peaks, which have been discussed in the literature [17].

PMMA also has characteristic SIMS spectra (Figure 8(c)), similar to those described in the literature [18]: fragments ($C_5H_7O_2^+$ and $C_5H_9O_2^+$ at m/z =99 and 101, respectively) associated with the MMA repeat unit ($C_5H_8O_2$) appear protonated or result from the loss or addition of one hydrogen atom. Other oxygen-containing peaks (as C_4H_5O+, $C_7H_9O^+$, at m/z=69 and 109) are characteristic of the main chain with the pendant group. The methyl ester pendant group has also characteristic peaks at m/z 15 (CH_3^+) and 59 ($C_2H_3O_2^+$). Branched hydrocarbon ions observed in the PMMA SIMS spectrum ($C_4H_5^+$, $C_4H_7^+$,$C_5H_7^+$, $C_5H_9^+$, $C_6H_7^+$,$C_6H_9^+$, $C_7H_9^+$, $C_7H_{11}^+$, and $C_8H_{11}^+$ at *m/z* 53, 55, 67,69, 79, 81, 93, 95, and 107, respectively) can be directly related to the PMMA backbone fragmentation [18-19].

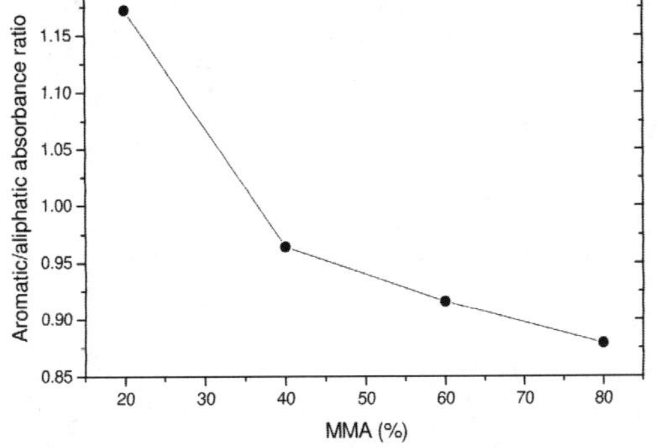

Figure 7 FTIR aromatic vs. aliphatic C-H stretching absorbance ratio at variable MMA fraction.

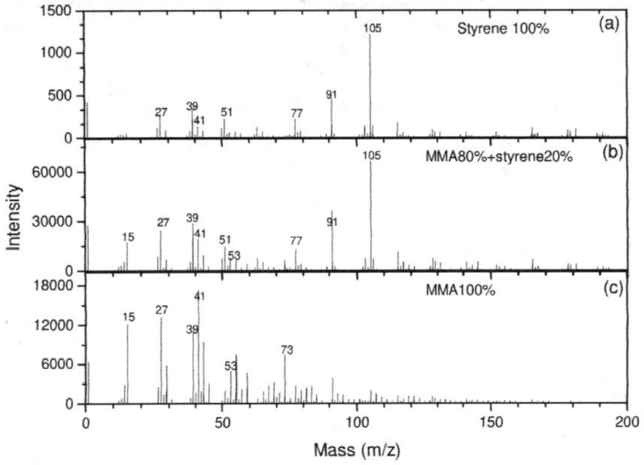

Figure 8 Positive SIMS spectra of (a) polyStyrene, (b) a copolymer, and (c) pMMA.

The SIMS spectrum of a copolymer containing 20% styrene is shown in Figure 8(b). Even with only an addition of 20% styrene in the copolymer, the polystyrene characteristic peaks dominated the

SIMS spectra, compared with most peaks of the MMA. The absolute intensity of peaks from styrene or MMA units was monitored as a function of the MMA content: hydrocarbon fragments such as CH_3^+, $C_2H_3^+$, $C_2H_5^+$, $C_5H_5^+$, and $C_7H_9^+$ at *m/z* = 15, 27, 29, 41, and 93, respectively, decreased with increasing styrene content,. By contrast, some characteristic styrene peaks (such as $C_5H_3^+$, $C_7H_7^+$, $C_8H_7^+$, and $C_8H_9^+$ at *m/z* = 63, 91, 103, and 105, respectively) showed an absolute intensity higher than those observed for polystyrene. In the SIMS spectra of copolymer and homopolymer films (Figure 8), the intensity of the fragments containing oxygen became very weak. The $C_7H_7^+$ tropylium ion is characteristic of the styrene unit, but we found its intensity does not quickly increase with the addition of styrene monomer in the plasma gas.

Figure 9 shows the intensity of fragments from the PMMA, such as CH_3^+, $C_3H_5^+$, and $C_3H_3O^+$: a large increase in intensity with the MMA content is observed. Several aromatic fragments from the copolymers, related to the styrene unit, like $C_6H_5^+$, $C_7H_7^+$, and $C_8H_9^+$, are evolving too (Figure 10). All these characteristic fragments exhibit higher peak intensities for the copolymer films, and are even higher than those observed for homopolymer films.

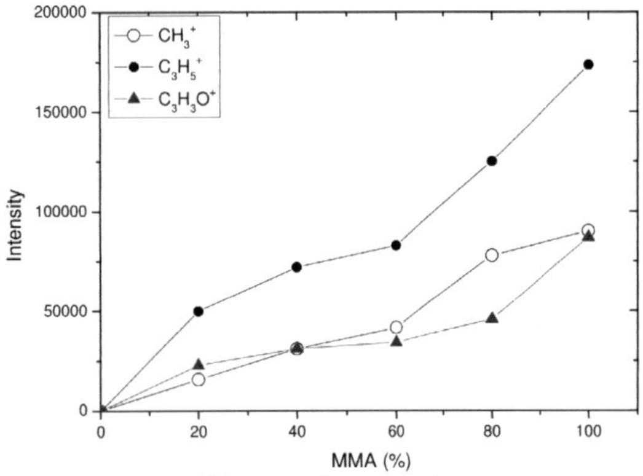

Figure 9 CH_3^+, $C_3H_5^+$ $C_3H_3O^+$ as a function of MMA content in styrene and MMA copolymer.

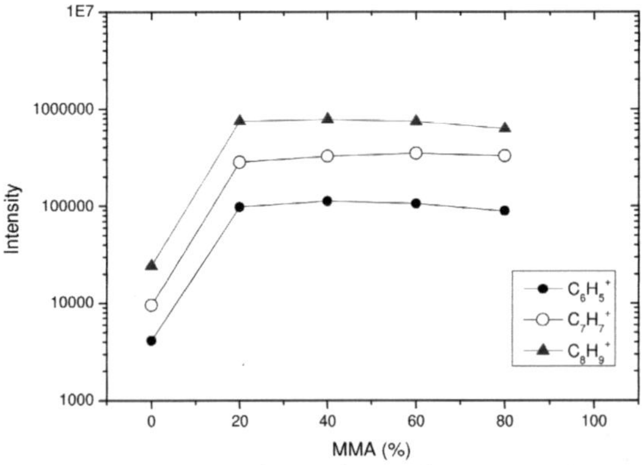

Figure 10 $C_6H_5^+$ $C_7H_7^+$ $C_8H_9^+$ as a function of MMA content in styrene and MMA copolymer.

5. Conclusion

By controlling the monomer content in the gas phase, plasma copolymerization can be used to deposit polymer thin films with different composition. This is due to reactions in the gas phase that depend strongly on the bond dissociation of each monomer and the reaction among the species in plasma gas. Some typical functional groups from styrene and MMA, which were identified in the OES, can be directly found in the copolymer films. Both FT-IR and XPS show that the contribution from the styrene monomer is very strong in the deposited films. The OES spectra also showed that the characteristics from styrene dominated the OES spectra of the co-monomer plasma. The π-π^* C1s peak (characteristic of aromatic rings) in the films increases quickly while the amount of C_6H_6 benzene excited species in the OES increases. This is confirmed from SIMS spectra.

Acknowledgments

We thank the financial support of Inter-University Attraction Pole Programme (Belgium Science Policy, Contract P6-08).

References

[1]Yasuda H. et Hirotsu T. J. 1977 *J. Polymer Sci.: Poly Chem. Ed.* 15, 2749.
[2]Sakata Jiro, Yamamoto Minoru and Tajima Ichiro 1988 *J. poly. Sci. Part B: poly. Phys.* 26. 1721-1731.
[3] Simionescu Cristofor I., Simionescu Bogdan C. and Ioan, Silivia. 1985 *J. Macromolecular sci., Part A.* 22, 765-778.

[4] Dilks A. and Kay E. 1980 *Macromolecules.* 14 855.
[5] Biederman H. 1987 *Vacuum.* 37 367.
[6] Hochart F., Levalois-Mitjaville J. and De Jaeger R. 2000 *Poly.* 41 3159-3165
[7] Kranarian G. and Celanese Hoechst 2001*Opt. Eng.* 40 1024
[8] Li Zhiling, Gillon X., Houssiau L. and Pireaux J.-J.,. 2009 19th Int. Symp. Plasma chemistry(ISPC, 2009) Bochum, Germany
[9] Kurucz R. L. http://www.cfa.harvard.edu/amp/ampdata/kurucz23/sekur.html. *Kurucz Atomic Line Database* .
[10] Herzberg G. 1950 *Molecular spectra and molecular structure I. spectrum of diatomic molecules.* New York : D. Van Nostrand Company
[11] Herzberg G. *Molecular* 1966 *spectra and molecular structure III. Electronic spectra and electronic structure of polyatomic molecules.* D. Van Nostrand Company
[12] Pearse R.W.B. and Gaydon A.G. 1963 *The idendification of molecular spectra.* Chapman & Hall LTD
[13] Tanabashi A., Hirao T. and Amano T. 2007 *The astrophysical joural supplement series* 169 472-484
[14] Heintze M, Magureanu M. and Kettlitz, M. 2002 *J. appl. Phys.* 92 7022-7031
[15] Crosswhite H. M. 1972 *The hydrogen molecule wavelength table of Gerhard Heinrich Dieke.* New York : Wiley
[16] Rond C., and al. 2008. *Chem. phys.* 345 16-26
[17] Leggett Graham J., Ratner Buddy D. and Vickerman John C. 1995 *Surf. interface anal.* 23 22-28
[18] Briggs D., Fletcher Ian W. and Gonçalves Nancy M. 2000 *Surf. interface anal.* 29 303-309
[19] Briggs D. 1990 *Surf. interface anal.* 15 734-738

Journal of Physics: Conference Series **275** (2011) 012021 doi:10.1088/1742-6596/275/1/012021

Plasma chemistry modeling for an inductively coupled plasma used for the growth of carbon nanotubes

Ming Mao and Annemie Bogaerts
Research group PLASMANT, Department of Chemistry, University of Antwerp
Universiteitsplein 1, B-2610 Wilrijk-Antwerp, Belgium
E-mail: annemie.bogaerts@ua.ac.be

Abstract

A hybrid model, called the hybrid plasma equipment model (HPEM), is used to describe the plasma chemistry in an inductively coupled plasma, operating in a gas mixture of C_2H_2 with either H_2 or NH_3, as typically used for carbon nanotube (CNT) growth. Two-dimensional profiles of power density, electron temperature and density, gas temperature, and densities of some plasma species are plotted and analyzed. Besides, the fluxes of the various plasma species towards the substrate (where the CNTs can be grown), as well as the decomposition rates of the feedstock gases (C_2H_2, NH_3 and H_2), are calculated as a function of the C_2H_2 fraction in both gas mixtures.

1. Introduction

Carbon nanotubes (CNTs) are gaining increasing interest, due to their unique physical, chemical and electronic properties, giving rise to a variety of (potential) applications, including nanoelectronics, hydrogen storage and field emission devices [1-3]. CNTs can be produced by arc discharges [4], laser ablation [5], chemical vapor deposition (CVD) [6], as well as by plasma enhanced CVD (PE-CVD) [7-25], which appears to be a very promising technology for the synthesis of vertically aligned CNTs, of interest for field emission devices [7]. Several plasma sources have been applied already for the growth of CNTs, including direct current (dc) glow discharges [8-12], capacitively coupled (cc) radiofrequency (rf) discharges [13-15], inductively coupled plasmas (ICPs) [16-22] and microwave (MW) discharges [23-25]. The feedstock gases most often used are CH_4 or C_2H_2, mixed with either H_2 or NH_3 as so-called etchant gases, i.e., to produce "clean" CNTs, with a limited fraction of amorphous phases.

In this paper, we present results of plasma chemistry modeling performed for an ICP, operating in a mixture of C_2H_2 with either H_2 or NH_3, in order to investigate the role of the various plasma species created in these gas mixtures, for the CNT growth. The effect of the C_2H_2 fraction in both gas mixtures will be analyzed, and the influence of the two different etchant gases (H_2 or NH_3) will be compared.

2. Model description

The model used in this work is the so-called hybrid plasma equipment model, developed by Kushner and coworkers [26]. It consists of three main modules, i.e., (i) an electromagnetics module, which calculates the electric and magnetic fields in the reactor, (ii) an electron energy transport module, which simulates the behavior of the (fast) electrons by a Monte Carlo procedure or the Boltzmann equation, and (iii) a fluid kinetics simulation, which treats the other plasma species by conservation equations. A full explanation about this hybrid model can be found in [26].

The focus of the present calculations is on the plasma chemistry. Therefore, a large number of different plasma species have to be taken into account, including electrons, various ions, molecules and radicals, which can react with each other in various collision processes. The plasma species and their reactions considered in the model depend on the gas mixture under study. In [27] we have investigated the detailed plasma chemistry in CH_4/H_2, CH_4/NH_3, C_2H_2/H_2 and C_2H_2/NH_3 gas mixtures. The different species taken into account in the model, as well as their reactions, were tabulated in detail [27].

Table 1: Overview of the species taken into account in the C_2H_2/H_2 model

Molecules	Radicals	Ions	Electrons
H_2	H	H^+, H_2^+	e^-
C_2H_2	C_2H, C_2H_3, C_2, CH_2, CH, C	$C_2H_2^+$, C_2H^+, $C_2H_3^+$, C_2^+, CH_2^+, CH^+, C^+	
C_4H_2	C_4H, C_4H_3	$C_4H_2^+$, C_4H^+, $C_4H_3^+$	
C_6H_2, C_6H_4	C_6H, C_6H_3	$C_6H_2^+$, C_6H^+, $C_6H_4^+$	
C_8H_2, C_8H_4, C_8H_6	C_8H	$C_8H_2^+$, C_8H^+, $C_8H_4^+$, $C_8H_6^+$	
$C_{10}H_2$, $C_{10}H_6$	$C_{10}H$	$C_{10}H_6^+$	
$C_{12}H_2$, $C_{12}H_6$	$C_{12}H$	$C_{12}H_6^+$	

In the present paper, we focus only on the gas mixture of C_2H_2 with either H_2 or NH_3. Table 1 summarizes the plasma species included for the C_2H_2/H_2 chemistry, whereas Table 2 presents the extra species added when NH_3 is used as the dilution gas. In the C_2H_2/H_2 plasma, 31 electron impact reactions, 45 ion-neutral and 29 neutral-neutral reactions are taken into account, whereas an extra number of 43 electron impact reactions, 48 ion-neutral and 67 neutral-neutral reactions were added to the model when NH_3 is used as the dilution gas. All details about this plasma chemistry can be found in [27].

Table 2: Overview of the extra species taken into account when NH_3 is used as the dilution gas

Molecules	Radicals	Ions
NH_3	NH, NH_2	NH^+, NH_2^+, NH_3^+, NH_4^+
N_2, N_2H_2, N_2H_4	N, N*, N_2*, N_2H, N_2H_3	N_2^+, N^+
HCN	CN, H_2CN	HCN^+

3. Results and discussion

First some general calculation results will be presented for the C_2H_2/H_2 gas mixture, at a ratio of 20/80, a total gas pressure of 50 mTorr, and 100 sccm total gas flow rate. The other operating conditions are 300 W source power, 30 W bias power at the substrate electrode, both applied at an operating frequency of 13.56 MHz. The substrate is heated to 823 K. Next, the effect of the C_2H_2/H_2 gas ratio will be investigated and a comparison will be made with the C_2H_2/NH_3 gas mixture.

3.1. General calculation results for the C_2H_2/H_2 gas mixture

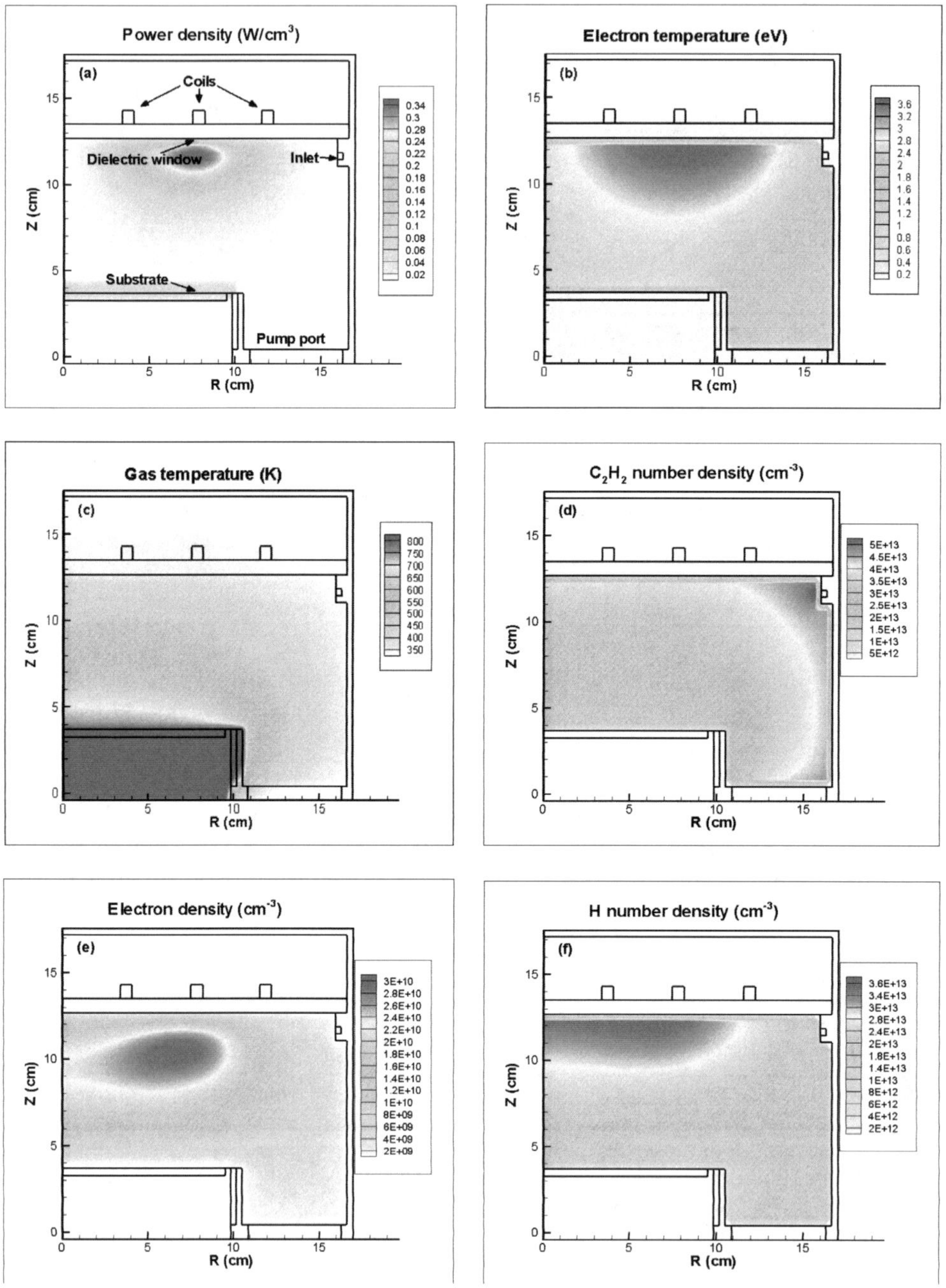

Figure 1: Calculated 2D profiles of (a) the power density, (b) the electron temperature, (c) the gas temperature, (d) the C_2H_2 gas density, (e) the electron density, and (f) the H atom density, in the C_2H_2 gas mixture at a ratio of 20/80, a total gas pressure of 50 mTorr, and a total gas flow rate of 100 sccm. The power applied at the coil and at the substrate is 300 W and 30 W, respectively, both with a frequency of 13.56 MHz. The substrate is heated to 823 K. Only one half plane of the cylindrically symmetrical reactor is shown. The coils, dielectric window, gas inlet, substrate and pump port are indicated in figure (a).

Figure 1 presents calculated 2D profiles of the power density (a), electron temperature (b), gas temperature (c), C_2H_2 gas density (d), electron density (e) and H density (f), in the ICP reactor under study. It is a transformer coupled plasma (TCP) reactor, with a flat coil at the top, which has been applied also experimentally for the growth of CNTs [16,18]. The reactor is cylindrically symmetric, so only one half plane of the reactor is shown. The coil, dielectric window, gas inlet, substrate and pump port are indicated in figure 1(a).

The power density (figure 1(a)) is at maximum just beneath the coil (i.e., the inductive region), and a second maximum is seen in front of the substrate electrode (i.e., the capacitive region). This is logical because the coil power is 300 W, whereas the power applied at the substrate electrode is only 30 W. As a result of this power input, the electrons in the plasma are heated to temperatures of a few eV, as is seen in figure 1(b). Again, the maximum is found beneath the coil. The gas temperature, on the other hand, is at maximum near the substrate, see figure 1(c). Indeed, the substrate is heated to 823 K to enable growth of CNTs [19-22], and the heat is spread out into the plasma. Without substrate heating, the gas would also heat to some extent, but only to about 350 K, as was demonstrated in [28]. Hence, the electron temperature is much higher than the gas temperature, as is typical for a non-LTE plasma.

As a result of this gas temperature profile, the C_2H_2 gas density reaches a minimum in front of the substrate, following the ideal gas law, as is shown in figure 1(d). Further, it is at maximum at the gas inlet, and drops to some extent in the reactor, as it disappears due to ionization and dissociation reactions. The electron density (figure 1(e)) reaches a maximum under the coil, similar to the electron temperature. This means that most of the electron impact reactions (ionization, excitation, dissociation,...) will take place in this region, and that most of the reactive species (i.e., ions, excited species, radicals) will also reach a maximum in this region. This is illustrated in figure 1(f) for the H atom density, but most of the other reactive plasma species indeed exhibit the same profile. The importance of the various reactive species will be discussed in the next section.

3.2. Effect of the C_2H_2/H_2 gas ratio

The carbon source for CNT growth is defined by the species impinging on the substrate and the decomposition activity on the catalyst surface. Therefore, the fluxes of the various plasma species arriving at the substrate surface are plotted in figure 2, as a function of the C_2H_2 fraction in the gas mixture. It is clear from figure 2(a) that H_2 is the dominant species at low C_2H_2 fraction, whereas C_2H_2 becomes more important at high C_2H_2 fraction. However, the turnover point is not seen at 50% C_2H_2, but at 80% C_2H_2, which indicates that H_2 is also formed to a large extent out of C_2H_2. Besides these two background gases, H atoms are also formed in large amounts, due to various reactions, such as dissociation and ionization of C_2H_2 and H_2, as is obvious from figure 2(b).

Other important neutral species bombarding the substrate, are $C_{2n}H_2$ (n=2, 3), C_6H_4 and $C_{2n}H_6$ (n=4, 5, 6) (see figure 2(a)), as well as the radicals $C_{2n}H_3$ (n=1, 2, 3) (see figure 2(b)).The fluxes of $C_{2n}H_2$ (n=2, 3) follow the same increasing trend as C_2H_2 upon increasing C_2H_2 gas fraction, while the C_6H_4 and $C_{2n}H_6$ (n=4, 5, 6) fluxes are less dependent on the C_2H_2 gas fraction. Indeed, the $C_{2n}H_2$ are mainly formed by C_2H insertion into $C_{2n-2}H_2$, which results in similar trends of the $C_{2n}H_2$ fluxes, whereas the

C_6H_4 and $C_{2n}H_6$ species are formed by neutralization of the corresponding ions on the wall. The H atom flux drops significantly with rising C_2H_2 fraction (and hence decreasing H_2 fraction), whereas the $C_{2n}H_3$ fluxes reach a maximum at about 40% C_2H_2. Indeed, these radicals are mainly produced by H insertion into $C_{2n}H_2$, and the decreasing trend of the H flux upon increasing C_2H_2 fraction, in combination with the increasing trend for the $C_{2n}H_2$ flux, results in a maximum at an intermediate C_2H_2 fraction of 40%.

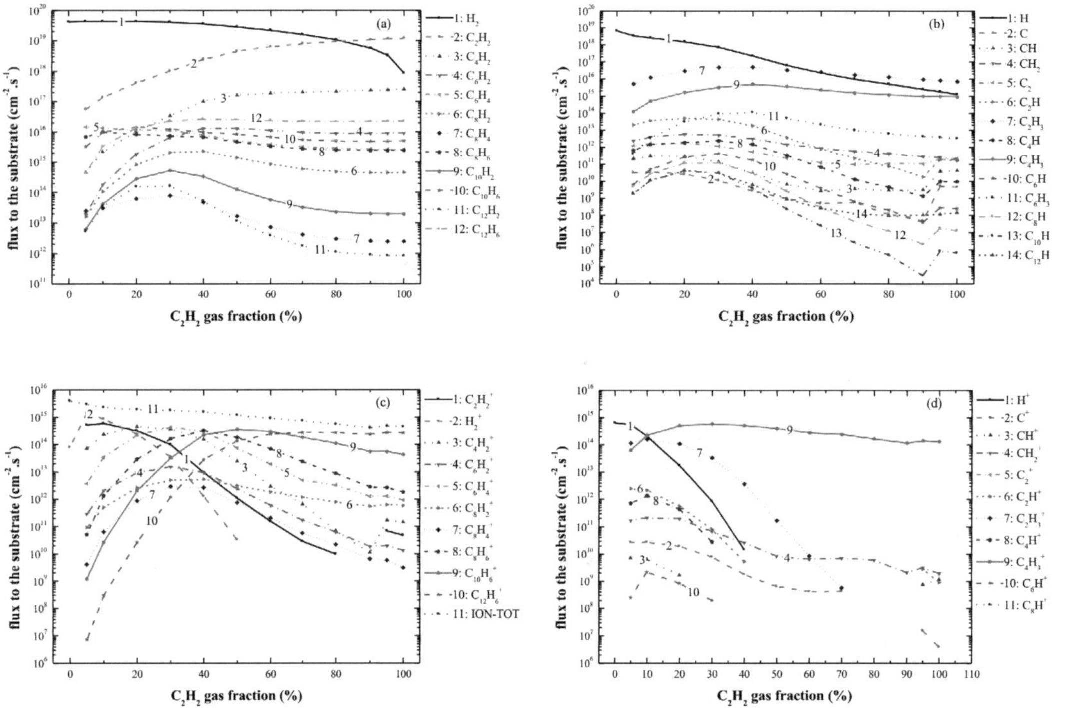

Figure 2: Calculated radially-averaged fluxes of the various molecules (a), radicals (b) and ions (c,d) bombarding the substrate, as a function of the C_2H_2 gas fraction, for the C_2H_2/H_2 gas mixture. The other operating conditions are the same as in figure 1.

With respect to the ions, it is clear from figure 2(c,d) that the lighter ions, such as H^+, H_2^+, $C_2H_2^+$ and $C_2H_3^+$, are predominant under H_2-rich conditions, whereas the heavier ions, such as $C_{10}H_6^+$ and $C_{12}H_6^+$, turn to be important under C_2H_2–rich conditions. $C_4H_3^+$ is characterized by a high flux for all gas fractions (see figure 2(d)), suggesting that this ion might be important in the CNT growth for the conditions under study. It is stated that hydrocarbon ions play a key role in the formation of amorphous carbon films [29]. Therefore, the fact that the larger hydrocarbon ions are predominant under C_2H_2-rich conditions might explain why these conditions are not so often used for CNT growth, and that most experiments are performed for a C_2H_2 fraction below 20% (e.g., [8,9,20]).

Finally, it is worth to mention that atomic carbon is hardly formed in the plasma (see figure 2(b)). Therefore, we can conclude that the carbon sources for CNT growth arise mainly from the decomposition of hydrocarbon molecules and radicals on the catalyst surface, as is also stated in [30].

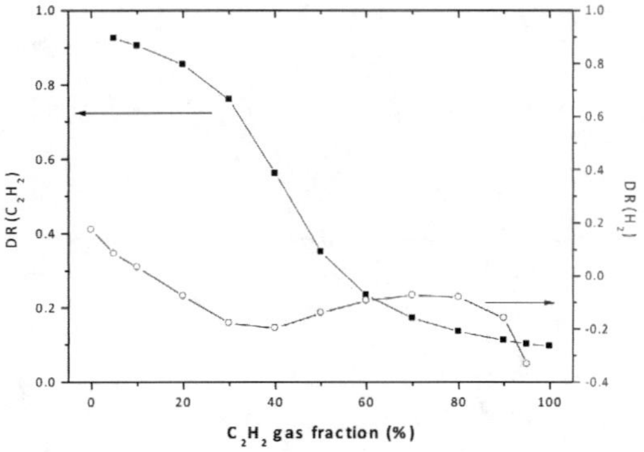

Figure 3: Calculated decomposition rates (DR) of the feedstock gas molecules (C_2H_2 and H_2), as a function of the C_2H_2 gas fraction, for the C_2H_2/H_2 gas mixture. The other operating conditions are the same as in figure 1.

Figure 3 illustrates the calculated decomposition rate (DR) of both background gases (C_2H_2 and H_2) in the plasma, as a function of C_2H_2 fraction. The DR is defined as follows:

$$DR=1-\frac{\text{volume-averaged density of background gas}}{\text{density of background gas at inlet}}$$

Hence, a DR of 1 indicates 100% decomposition, or conversion into other species. C_2H_2 has a high DR at low C_2H_2 gas fraction (see left axis), and the DR drops significantly upon increasing C_2H_2 fraction, especially in the range between 30 and 70%. The C_2H_2 species are believed to be the main precursors for CNT growth in a PE-CVD system [10,12]. Hence, a high DR of C_2H_2, in combination with a low C_2H_2 fraction in the gas mixture, gives rise to a low C_2H_2 flux to the substrate, which might suggest a relatively low (and hence: controlled) growth rate for the CNTs. This is indeed observed experimentally [9,20]. The DR of H_2 is found to be very low at low C_2H_2 fraction, and it is even negative at C_2H_2-rich conditions, as is obvious from figure 3 (right axis). These negative values for the DR indicate that more H_2 molecules are formed, e.g., out of C_2H_2, rather than decomposed.

3.3. Comparison of H_2 and NH_3 as dilution gases

When NH_3 is used as the dilution gas instead of H_2, some new species appear in the plasma, as is clear from Table 2 above. Figure 4 illustrates the fluxes of these extra species towards the substrate, as a function of the C_2H_2 gas fraction. N_2, N_2H_2 and NH_4^+ appear to be important neutrals and ions, respectively, besides the background gases C_2H_2 and NH_3. It was reported [25] that the presence of N_2 can enhance carbon diffusion into the catalyst, promoting CNT growth. Hence, this might explain why in many papers (e.g., [8-10,12,24]) NH_3 is used as the dilution gas, rather than H_2.

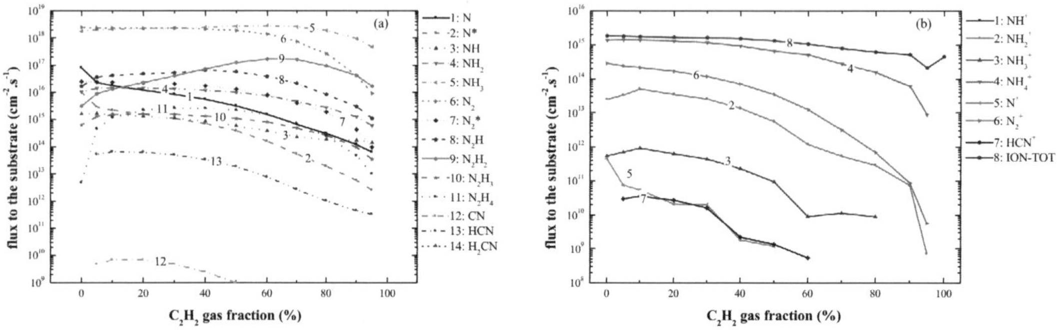

Figure 4: Calculated radially-averaged fluxes of the extra neutral species (a) and ions (b) bombarding the substrate, as a function of the C_2H_2 gas fraction, for the C_2H_2/NH_3 gas mixture. The other operating conditions are the same as in figure 1.

The DRs of C_2H_2 and NH_3 are plotted against C_2H_2 fraction in figure 5. The DR of C_2H_2 shows the same trend as in the C_2H_2/H_2 gas mixture (cf. figure 3 above). Moreover, the DR of NH_3 drops also from nearly 1 at low C_2H_2 gas fraction to values around 0.4 at high C_2H_2 fraction. This leads to a pronounced drop in the atomic H production. It is generally known that H can etch away amorphous carbon phases formed in the growing CNTs [20]. Therefore, the large flux of H atoms at low C_2H_2 gas fraction, in combination with the low fluxes of hydrocarbon ions towards the substrate, suggests that "clean" CNT growth conditions can be achieved at a low C_2H_2 gas fraction. This is also reported in literature. Indeed, Chhowalla et al. [8] found that CNTs could be grown at a C_2H_2 fraction between 5 and 30%. Also in [24] the longest CNTs with the highest density were obtained for the smallest C_2H_2/NH_3 gas ratio (15/150), corresponding to around 10% of C_2H_2.

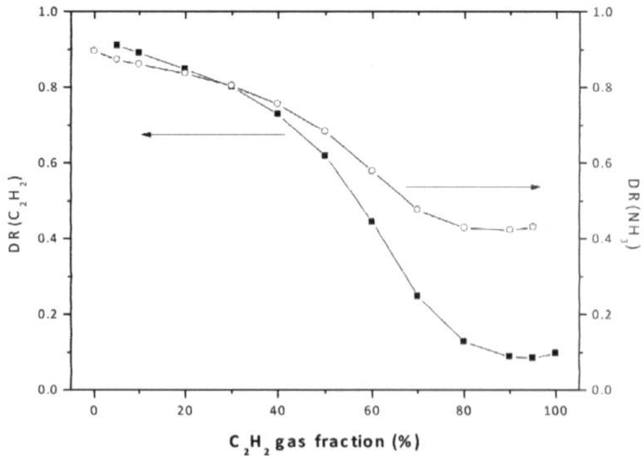

Figure 5: Calculated decomposition rates (DR) of the feedstock gas molecules (C_2H_2 and NH_3), as a function of the C_2H_2 gas fraction, for the C_2H_2/NH_3 gas mixture. The other operating conditions are the same as in figure 1.

5. Conclusions

We have applied a hybrid model to investigate the plasma chemistry in an ICP reactor in C_2H_2/H_2 and C_2H_2/NH_3 gas mixtures. First, some general calculation results were illustrated, such as the power density, electron and gas temperature, electron density and densities of some other plasma species. Subsequently, the fluxes of the various plasma species bombarding the substrate, as well as the decomposition rates (DRs) of the feedstock gases were analyzed for different C_2H_2/H_2 and C_2H_2/NH_3 gas ratios.

It is found that both gas mixtures give rise to a very rich plasma chemistry, with several large hydrocarbon molecules, radicals and ions, especially at a high C_2H_2 gas fraction. When NH_3 is applied as the dilution gas, N_2, N_2H and N_2H_2, as well as NH_4^+, are also important plasma species. The C flux, on the other hand, was found to be the lowest of all radicals, which confirms that the carbon source for CNT growth mainly arises from the decomposition of hydrocarbon molecules on the catalyst surface [30].

The calculated DRs of the feedstock gas molecules illustrate that C_2H_2 is almost fully decomposed at low C_2H_2 fraction in both gas mixtures, but the DR drops significantly upon higher C_2H_2 gas fraction. The DR of H_2 was found to be very low, or even negative at a C_2H_2 fraction above 10%, indicating that more H_2 molecules are formed (out of e.g., C_2H_2), rather than decomposed. The DR of NH_3 follows a similar trend as the C_2H_2 DR, being high at low C_2H_2 fraction, and decreasing for higher C_2H_2 fractions. This results in a drop of the atomic H flux towards the substrate (or the growing CNT) at high C_2H_2 fraction. Moreover, the higher hydrocarbon fluxes also exhibit a maximum at these conditions. Therefore, we expect that an amorphous carbon film will be formed under these conditions, which cannot easily be etched away, because of the low amount of etching H atoms, and this will suppress the growth of CNTs. Hence, we expect that a low fraction of C_2H_2 and hence a higher fraction of H_2 or NH_3 in the gas mixture will result in more "clean" conditions for CNT growth. Indeed, experiments reported in literature appear to be performed typically at a C_2H_2 fraction below 20% (e.g., [8,9,20]).

Acknowledgements

The Fund for Scientific Research (FWO Flanders) and the Prime Minister's Office through IAP-VI are acknowledged for financial support. The calculations were performed on the CalcUA computing facilities of the University of Antwerp. We are also very grateful to M. Kushner and group members from providing the HPEM and useful advice.

References

[1] Tseng G Y and Ellenbogen J C 2001 *Science* **294** 1293-4.
[2] Darkrim F L, Malbrunot P and Tartaglia G P 2002 *Int J Hydrogen Energ* **27** 193-202.
[3] Guillorn M A, Melechko A V, Merkulov V I, Ellis E D, Simpson M L, Baylor L R and Bordonaro G J 2001 *J Vac Sci Technol B* **19** 2598-601.
[4] Journet C, Maser W K, Bernier P, Loiseau A, delaChapelle M L, Lefrant S, Deniard P, Lee R and Fischer J E 1997 *Nature* **388** 756-8.
[5] Yudasaka M, Komatsu T, Ichihashi T and Iijima S 1997 *Chemical Physics Letters* **278** 102-6.
[6] Lee C J, Kim D W, Lee T J, Choi Y C, Park Y S, Lee Y H, Choi W B, Lee N S, Park G S and Kim J M 1999 *Chemical Physics Letters* **312** 461-8.
[7] Cassell A M, Ye Q, Cruden B A, Li J, Sarrazin P C, Ng H T, Han J and Meyyappan M 2004 *Nanotechnology* **15** 9-15.

[8] Chhowalla M, Teo K B K, Ducati C, Rupesinghe N L, Amaratunga G A J, Ferrari A C, Roy D, Robertson J and Milne W I 2001 *J Appl Phys* **90** 5308-17.
[9] Bell M S, Lacerda R G, Teo K B K, Rupesinghe N L, Amaratunga G A J, Milne W I and Chhowalla M 2004 *Appl Phys Lett* **85** 1137-9.
[10] Bell M S, Teo K B K and Milne W I 2007 *J Phys D Appl Phys* **40** 2285-92.
[11] Hash D, Bose D, Govindan T R and Meyyappan M 2003 *J Appl Phys* **93** 6284-90.
[12] Bell M S, Teo K B K, Lacerda R G, Milne W I, Hash D B and Meyyappan M 2006 *Pure Appl Chem* **78** 1117-25.
[13] Okita A, Suda Y, Ozeki A, Sugawara H, Sakai Y, Oda A and Nakamura J 2006 *J Appl Phys* **99** 0143021-7.
[14] Okita A, Suda Y, Oda A, Nakamura J, Ozeki A, Bhattacharyya K, Sugawara H and Sakai Y 2007 *Carbon* **45** 1518-26.
[15] Oda A, Suda Y and Okita A 2008 *Thin Solid Films* **516** 6570-4.
[16] Delzeit L, McAninch I, Cruden B A, Hash D, Chen B, Han J and Meyyappan M 2002 *J Appl Phys* **91** 6027-33.
[17] Meyyappan M, Delzeit L, Cassell A and Hash D 2003 *Plasma Sources Sci T* **12** 205-16.
[18] Cruden B A and Meyyappan M 2005 *J Appl Phys* **97**-84311-15.
[19] Wei H W 2008 Study of Growth of Vertically-Aligned Carbon Nanofibers by Plasma Enhanced Chemical Vapor Deposition-Growth Mechanism and Field Emission. In: *Department of Engineering and System Science,* (Hsinchu, Taiwan: National Tsing Hua University) p 175.
[20] Lin Y Y, Wei H W, Leou K C, Lin H, Tung C H, Wei M T, Lin C and Tsai C H 2006 *J Vac Sci Technol B* **24** 97-103.
[21] Wei H W, Leou K C, Wei M T, Lin Y Y and Tsai C H 2005 *J Appl Phys* **98** 044313.
[22] Yang C S 2005 Low temperature growth of single-walled carbon nanotubes by PECVD. In: *Department of Engineering and System Science,* (Hsinchu, Taiwan: National Tsing Hua University) p 88.
[23] Maschmann M R, Amama P B, Goyal A, Iqbal Z, Gat R, Fisher T S, 2006 *Carbon* **44** 10-18.
[24] Wang P, Lu J and Zhou O 2008 *Nanotechnology* **19** 185605-1-7.
[25] Lin C H, Chang H L, Hsu C M, Lo A Y and Kuo C T 2003 *Diam Relat Mater* **12** 1851-7.
[26] Ventzek P L G, Sommerer T J, Hoekstra R J and Kushner M J 1993 *Appl Phys Lett* **63** 605-7.
[27] Mao M and Bogaerts A 2010 *J Phys D Appl Phys* **43** 205201-1-20.
[28] Mao M and Bogaerts A *J Phys D Appl Phys* (submitted).
[29] Grigonis A, Sablinskas V, Silinskas M and Tribandis D 2004 *Vacuum* **75** 261-7.
[30] Hofmann S, Sharma R, Ducati C, Du G, Mattevi C, Cepek C, Cantoro M, Pisana S, Parvez A, Cervantes-Sodi F, Ferrari A C, Dunin-Borkowski R, Lizzit S, Petaccia L, Goldoni A and Robertson J 2007 *Nano Lett* **7** 602-8.

AUTHOR INDEX